Reviewing Britain's Defence

IAN BELLANY
Lancaster University

Dartmouth
Aldershot . Brookfield USA . Hong Kong . Singapore . Sydney

Published by
Dartmouth Publishing Company Limited
Gower House
Croft Road
Aldershot
Hants GU11 3HR
England

Dartmouth Publishing Company
Old Post Road
Brookfield
Vermont 05036
USA

British Library Cataloguing in Publication Data
Bellany, Ian
Reviewing Britain's Defence
I. Title
355.033541

Library of Congress Cataloging-in-Publication Data
Bellany, Ian
Reviewing Britain's defence / Ian Bellany.
p. cm.
Includes bibliographical references and index.
ISBN 1-85521-462-8 : $59.95
1. Great Britain--Military policy. 2. Defense industries--Great Britain. I. Title.
UA647.B42 1994
355'.0335'41--dc2 93-40733
CIP

ISBN 1 85521 462 8

Printed and Bound in Great Britain by
Athenaeum Press Ltd, Newcastle upon Tyne.

Contents

Preface vii

Introduction 1

Section One: Framework for Analysis 5

Section Two: The Army, the Navy and the Air Force 45

Section Three: The Defence-Contracting Industries 88

Appendix: UK R & D 1989 154

Conclusion 157

Annex: The Size of the British Defence Budget 167

Appendix 1: Regression Analysis 188

Appendix 2: Game Theory 196

Index 202

Preface

The origins of this book lie in a series of lectures the author gave on the Royal Navy and British defence to the Britannia Royal Naval College at Dartmouth over a number of years in the 1970s and early 1980s, as one of a visiting team of academics and others annually assembled and headed by Admiral Sir James Eberle.

There are only very faint traces of the actual substance of these lectures in this book. This is not so much because they failed to survive Admiral Eberle's jovial scepticism, but more because the lectures became in the end, unintentionally, the author's way of learning something about the navy and defence policy, at the public's expense. The lectures nonetheless all reflected the author's puzzlement at what the navy was for in the nuclear and post-imperial age, and each one was an ultimately unsatisfactory attempt at coming up with an answer to Duncan Sandys's original penetrating enquiry.

Various possible explanations were essayed in turn. 'Standing ready to fight a third Battle of the Atlantic' did not really work. For one thing, a prolonged struggle to protect the NATO reinforcement routes, which could more or less plausibly be made to justify a big navy, as the British navy was, would have suggested an official zeal for the then comparatively recently agreed NATO strategy of 'flexible response' more to be associated with the 'whiz kids' of McNamara's Pentagon than with Whitehall. Anyway, whilst a flexible response navy might than with Whitehall. Anyway, whilst a flexible response navy might reasonably have been as large and as expensive as the Royal Navy was, it would surely have contained different sorts of ships. Convoying - the one lesson of naval history that has apparently to be re-learnt with every fresh campaign being that convoying is the only reliable method for protecting reinforcement routes - means a requirement for small escort vessels and a lot of them. And these are what a flexible response navy would need. The Royal Navy, on the other hand, did not seem greatly to care for such ships.

The second explanation was unsuccessful in a different way. Britain is a trading country and trade is important in the sense that the British economy proportionately depends on international trade to a greater degree than almost any other. Trade, or most of it, continues to come by sea, and protecting trade routes against disruption, perhaps as a result of comparatively small-scale regional conflicts affecting some 'choke point', might itself call for a large British navy. There could easily be more than one regional conflict going on at once, and maintaining even a small number of vessels permanently on station in some corner of the globe is subject to an inverse multiplier effect. Each ship on station ties up at least one, possibly two others, that are being got ready to relieve her on rotation. In the context of this kind of explanation of what the navy was for, the appropriate kind of navy would indeed be more general purpose in appearance, with large ships and small. The trouble with this account of what the post-1970 navy was for was the accession of Britain to the Common Market. This fact in itself merely made formal a reorientation of the direction of British trade which had been going on for years, away from the open sea and towards the Continent. The discovery and exploitation of British oil in the North Sea merely confirmed that the typical length of British trade routes had shrunk by the mid-1970s to something not very much more than that of the average ferry crossing between the east coast and the European mainland. This is not to say that even such short routes (and oil platforms) do not require naval protection; conceivably the amount of protection necessary might even have justified a large navy. The real difficulty again lay in the kind of navy it would have justified. The British mainland would now be an aircraft carrier, permanently moored at the only choke point that really mattered, not requiring to be relieved on rotation, provided free by nature, and, of course, unsinkable. A system of land-based aircraft, missilry and electronic detectors above, on and under the surface was all that the navy needed to provide. Of course, no such navy existed and the Admiralty was certainly not planning one.

The series of lectures at Dartmouth ended just after the Falklands War and not long before the author ran out of explanations of what the navy was for.

Ten long years of subsequent taking stock have resulted in the present book.

No part of the explanation for the delay can be laid at the feet of the author's employers. In fact Lancaster University was generous with financial support for the project, at a crucial time, from its social science research fund. Aside from what has come to be accepted as the normal vicissitudes of academic life, the chief reason for the delay is a simple one. It took years for the author to lose his naive belief that there was in real life likely to be more than the crudest correspondence between the kind of navy that was in existence at any one time and the kind of navy that seemed to be required in terms of the stated defence priorities of the government of the day. This realisation was accompanied by another which expanded the task in hand: that in order to understand the navy it was necessary to investigate more than the navy alone.

Whilst this book might just possibly have eventually been completed without the financial assistance referred to, it would never have been written at all without two sorts of intellectual assistance. The body of published material produced by British academic specialists on defence issues proved to be utterly indispensable. In his high regard for the rich diversity of this work, the author probably finds himself in disagreement with Sir John Nott, in spite of his admiration for the latter's general perspicacity on defence matters. Even more valuable was the assistance the author received from his hard-pressed but generous colleagues at Lancaster University and in particular its Department of Politics and International Relations. The Department and the University have long contained a selection of unusually independent-minded specialists on defence issues. Especially helpful, in their very different ways, were Bob Baker, Mick Dillon, Martin Edmonds and John Gooch. And without the ready access granted to the comprehensive stock of research materials on British defence maintained by Lancaster's Centre for Defence and International Security Studies, headed by Dr Edmonds, it is extremely unlikely that the sabbatical term at the author's disposal in 1991/92 would have been nearly enough to speed the project towards its end.

In addition, when the time came to turn the manuscript into a book, the author has done nothing to deserve an editorial team possessing such eagle-eyes and near-omniscience as that at Dartmouth Publishing.

Almost needless to say, the contributions from colleagues at Lancaster and elsewhere have been exclusively to that which the reader may find good in what follows; the rest is the author's responsibility alone.

Introduction

This book is an attempt to review and to understand British conventional defence policy in the twenty-year period since 1970, from approximately which date onwards British forces were meant to have been withdrawn from a global role.

It asks a number of specific questions about conventional defence policy. Why has the British defence budget remained so large compared to those, on average, of its European NATO allies? What is the purpose of the Royal Navy, itself still larger in tonnage than the combined strength of the French and German navies if, as is generally believed, nuclear weapons have all but ruled out a major conventional, attritional war between nuclear-armed powers? And why does Britain continue to maintain a 'defence industrial base' (domestic defence-contracting industries covering almost the entire spectrum of the material requirements of the armed services) when the classic rationale of the danger of a prolonged major war had virtually disappeared with the maturing of the East-West nuclear balance?

In order to answer these questions, it was found necessary to construct a framework for analysis. This is a free adaptation of the well-known successful framework for the analysis of United States' military nuclear policy usually associated with the names of Desmond Ball and Donald Snow.

At its core is the suggestion that there is not one but three levels of defence policy. One level is declaratory, or stated, policy. This is set out in regular government statements, most comprehensively in this country in the annual statement accompanying the budget estimates for defence. Another level is the operational, which corresponds to what the armed services actually think of themselves as being held in readiness to do and being prepared for. The third level resembles the second, but corresponds to what the defence-contracting sector of the British economy regards itself as being for, and holds itself ready to do; this is called here the development-industrial level. In an ideal world these three levels of policy would be mutually coherent or 'in line' and there would be no contradictions between them. In the real world the reverse is more usually the case, and it is within these contradictions that the answers to our questions lie.

In addition, or, rather, complementarily, a more direct attack, outside the framework, on one of the above questions - that concerning the size of the British defence budget - was also found to be productive. This is contained in an annex, which would anyway have been its natural home even if it had fitted squarely into the chosen analytical framework, because it adopts an approach based on statistical analysis. The results of this analysis are compatible and intersect with those that emerge from the main body of the work. But there is also an unexpected finding which is not explicitly echoed elsewhere in the text. This is the possibility that some part of the apparent excess size of the British defence budget compared to those of its European allies may be due to something that hitherto has been seen as afflicting the Americans alone - 'free riding' on the British defence budget by a number of its smaller allies in NATO. In other words, some part of the excess size of the British defence budget in a sense does not really need to be explained at all: it is a mirage created by a tendency of some of its allies to spend less on defence than otherwise might have seemed appropriate.

The framework for analysis also serves as an organising principle for the book. The text, apart from the annex, is divided into three parts. The first part introduces the framework for analysis and attempts a clarification of what may be meant by the declaratory level of policy, as well as giving an account of stated policy since 1970. Evidence for the latter presents no difficulty - it is simply what government states defence policy to be. Oddly enough, whilst the declaratory level is the most open of the three, it is also the most difficult to define in terms of the classes of persons who actually operate at this level.

The second part of the book concerns the operational level. Two approaches to the discovery of possible contradictions between this and the stated level are possible. One is direct. Since 1970 Britain has fought in two wars, in the Falklands in 1982 and in the Gulf against Iraq (1991/92), or three if the secret yet much longer low-intensity war in Oman between 1965 and 1975 is taken into account. These wars demonstrated what British forces were in practice capable of doing and are directly revealing of the nature of the operational level. In addition, clues as to contradictions between levels can be found in the existence of contradictions within levels. Senior serving officers are by convention allowed to operate at the declaratory level from time to time. They publish occasional articles and give lectures to public forums. It is among these statements that contradictions are most often found. For instance, a senior naval spokesman indicating the navy's readiness for a comparatively long major war in Europe - lasting weeks and possibly months - can be compared with a senior airman's saying to the same forum, exactly a fortnight before, that the RAF is ready for the same war, provided it lasts only days or weeks at the most. Luckily, a single forum exists where a conveniently large sample of such statements can be contrasted on a comparable basis. This is the Royal United Services Institute (RUSI) for Defence Studies in London, and whilst there are other, similarly useful forums for our purpose - the Royal College for Defence Studies, for instance - it is doubtful if this book would have been possible without the excellent reports provided by the Journal of the RUSI.

The third part of the book concerns the development-industrial level. The investigative approach taken to this level of policy roughly parallels that taken to the operational. In place of actual campaigns there are a number of interested and disinterested factual and internationally comparative accounts of the size and scope of the British defence-contracting industry, including good statistical data provided by government.

When governments state that their philosophy of purchasing equipment for the armed services is 'value for money', and then continue to make these purchases (procurement) almost exclusively within Britain, if data on the British defence-contracting industry indicate that it does not always provide the best value for money, there is a direct contradiction. And there are other contradictions. Senior spokesmen for the development-industrial level have more freedom than their military counterparts to speak their minds in public. When the latter do pronounce in public, they are essentially constrained to operate at the declaratory level. At most, 'coded' language may sometimes be used, so that what is said may carry a different meaning to operational than to declaratory listeners. When spokesmen for the defence-contracting industries speak, they are more often speaking for themselves, that is to say the development-industrial level, although they often share platforms, such as the Royal United Services Institute, with the operational level. This plainer speaking results in a corresponding increase in the blatancy of contradictions both between and within levels. Thus, the development-industrial level has its own private strategic rationale for the maintenance of a British 'defence industrial base' - namely, that allies can never be trusted to supply armaments of the right sort at the right time. This of course is intended to imply that foreign purchases of military equipment should therefore in effect never be contemplated. Since 1979 at least, this has been in flat contradiction to what has been said at the declaratory level about procurement policy. In addition to which, the fact that Britain since the early 1960s has been wholly dependent upon an ally and a single ally at that - the United States - for key components of its strategic nuclear force, shows that contradictions within levels are not a monopoly of the declaratory.

The concluding section of the book contains answers to the questions posed at the beginning. In an ideal world, perhaps, this is all that it should contain, but in practice it is more extensive. It contains new questions raised by the attempt to answer those we began with: it also contains answers to additional questions that had not originally been raised at all, in particular questions relating to the nature of that remarkable British political phenomenon known as the 'defence review'.

Section One: Framework for Analysis

Introduction

Students of the strategic nuclear policy of the United States are nowadays quite ready to distinguish between what Desmond Ball calls different 'elements' and Donald Snow different 'levels' of that policy. The Ball/Snow contribution to thinking about United States' nuclear strategy comes in their seminal (if only partly original) suggestion that there are three major and distinct ways in which that strategy or policy can be understood. According to Ball, writing in 1983:

> Strategic nuclear policy is composed of several different elements, of which the most important are declaratory policy, action or employment policy, and force development policy. Declaratory policy is that set of public pronouncements made by the President, the Secretary of Defense, or sometimes other senior administration officials, regarding the requirements of deterrence, targeting policy and strategic doctrine. Action or employment policy, on the other hand, comprises the actual war-fighting strategy that the US would adopt in a nuclear exchange, the designation of the forces to be used, the targets to be hit, the allocation of the forces to those targets, and the rate at which the exchange is to proceed.

> Force development policy is the guidance that informs decisions on such matters as the size and capabilities of the US ICBM force, the requirements for a new strategic bomber, and the size and characteristics of the SLBM force - i.e. the forces available for the implementation of the action policy in the event of a nuclear exchange. Some aspects of these disparate elements are inconsistent with others and, indeed, there are contradictions within as well as between some of the particular elements.
>
> Employment policy would logically seem to be the pre-eminent element of US strategic nuclear policy. Not only is it potentially by far the most consequential, but rational policy-making requires decisions as to *how* the US would actually use her strategic forces in the event of a nuclear exchange should precede decisions concerning the acquisition and deployment of those forces. Unfortunately, however, employment policy is also the least understood element of US strategic nuclear policy - by policy-makers no less than by the strategic community in general.[1]

Writing later, in 1987, Snow's mature conceptualisation of Ball's 'elements' (Snow's first contributions on the subject are dated 1983, which, unlike his writings in 1987, acknowledge earlier - 1981 - work by Ball on the same subject[2]) speaks of three (or perhaps four) 'levels' of nuclear strategy. One is the declaratory level:

> Declaratory strategy is the province of the president, his top defense advisers, and the 'defense intellectuals' in government and academia who think and write about nuclear matters. Declaratory strategy is ... strategy at its most abstract, consisting of theoretical speculations and postures. For this reason, it is the level of strategy to which academic intellectuals ... are most drawn and where they are most effective. Manifestations of declaratory strategy in places such as presidential speeches ... are often

1 Desmond Ball, 'Targeting for Strategic Deterrence', *Adelphi Paper*, No.185, International Institute for Strategic Studies, London, Summer 1983, p. 37.

2 Desmond Ball, 'Dynamics of the US Strategic Nuclear Policy Framework', *Defense and Foreign Affairs*, April 1981.

written or influenced by these individuals.[3]

Snow's second level of strategy is essentially Ball's third 'element', subdivided into two parts or 'phases': development and deployment strategy. This level:

> encompasses the period from the discovery of the weapons potential of some physical fact about the universe until that discovery results in an actually deployed weapons system or component. It is, in other words, the physical process from ideation to production of a weapon.[4]

According to Snow, the first phase of his second level is one of development of the weapon, or component of a weapon, from the first sighting on the technical horizon of an idea with warlike potential to the testing of a working prototype. This phase is 'dominated by scientists and engineers who proceed at whatever pace the discovery period allows'. The second phase is deployment strategy, covering the decision to purchase final versions of the prototype and the acceptance of the weapon into the inventory. The second phase, according to Snow, is 'dominated by politicians'.[5]

Snow notes that because the second level involves processes which can take a long time, 'the development and deployment decisions of one administration will be most strongly felt by subsequent administrations'.[6]

Snow's third level is Ball's second element, employment strategy or, as Ball sometimes calls it, 'action' strategy, which is for Snow 'the blueprint for carrying out the threats that declaratory strategy makes'. Snow elaborates:

> Employment strategy is the least visible to the public eye. The reason is that this level of strategy is made privately, and the actual nuclear war-fighting document (really a set of computer

---

3 Donald M. Snow, *National Security: Enduring Problems of U.S. Defense Policy*, St Martin's Press, New York, 1987, pp.189, 192.

4 *ibid.*

5 *ibid.*

6 *ibid.*

> programs) is among the most highly classified of government secrets, at least in its detailed version. The embodiment of employment strategy is something called the SIOP, or Single Integrated Operational Plan. The plan is the product of an inter-service group known as the Joint Strategic Targeting and Planning Staff (JSTPS), consisting of officers and civilians, that is an agency of the Joint Chiefs of Staff. The JSTPS began operation in the early 1960s and produces plans as directed by the president and the secretary of defense.
>
> Although the details of this process are heavily guarded, the general facts are publicly available ... the SIOP has changed incrementally rather than dramatically over time, with a consistent emphasis on counterforce targeting, regardless of what declaratory strategy might suggest.[7]

Snow, when writing earlier, in 1983, goes on to specify that military personnel dominate the 'operational element of employment strategy'[8], although by 1987 he did not find it necessary to specify who it was that 'manned' the employment level.

Other thinkers: the place of academics

United States

It should be evident that whilst the Ball/Snow analysis is directed at the nuclear strategic agenda of the United States, there is nothing about it that necessitates an exclusive focus on nuclear matters. The same or at least a very similar analysis could be carried out on the conventional military strategy of the United States although, as far as the present author is aware, this has never been explicitly attempted in a comprehensive way. However, a 1988 work by D.M. Shafer on United States' counterinsurgency policy adopts an analytical device which could be described as an enlarged version of the Ball/Snow approach to a smaller subset of United States' security concerns. Shafer's method

---

7 *ibid.*

8 Donald M. Snow, 'Levels of Strategy and American Strategic Nuclear Policy', *Air University Review*, Vol. XXXV, No. 1, Nov.-Dec. 1983, pp. 63, 73.

moreover makes more explicit the debt that the Ball/Snow approach might owe to earlier attempts to understand United States' foreign policy (embracing defence and security policy) from a perspective other than that of the 'unitary actor'. Shafer quotes I.M. Destler's 1979 description of two competing definitions of 'policy' as 'an authoritative statement ... prescribing general criteria for action', which we may see as equivalent to declaratory policy, and 'what government actually does', which would seem roughly to correspond to action or employment policy.[9]

In fact this dual characterisation of what perhaps should be understood by 'policy' has a considerable lineage. Writing in *Foreign Affairs* in 1956 and anticipating something of the Ball/Snow conceptualisation by about twenty-five years,[10] Paul H. Nitze divides policy into 'action' policy - 'general guide lines which we believe should and will in fact govern our actions in various contingencies' - and 'declaratory' policy - 'policy statements which have as their aim political and psychological effects'. Nitze's initial characterisation, like Destler's, does not seem to restrict itself to any particular field of policy, civil or military, domestic or international, and it is clear when he does home in on warlike issues that he would not wish to restrict the remit of 'action' and 'declaratory' policy to the sphere of nuclear weapons. But the burden of what Nitze has to say concerns the contemporary nuclear strategy of the United States and the author's discomfort with the then nuclear declaratory policy of massive retaliation.

Shafer's own study of United States' counterinsurgency policy adopts five of what he calls 'competing types of explanation', one of which he names presidential politics, which corresponds roughly to the Ball/Snow declaratory strategy, and another of which he names bureaucratic politics, following earlier analysts such as Allison and Kaplan. If we permit ourselves to reinterpret Ball and Snow and see the military as one sort of bureaucracy and, admittedly with a little more expenditure of imaginative energy, the scientists and engineers as another, we have a rough identification with the sources of the Ball/Snow employment and developmental strategies. That there is nothing far-fetched at least

---

9 Quoted in D. Michael Shafer, *Deadly Paradigms: The Failure of US Counterinsurgency Policy*, Princeton University Press, Princeton, 1988.

10 Paul H. Nitze, 'Atoms, Strategy and Policy', *Foreign Affairs*, Vol. 34, No. 2, January 1956, p. 187.

about the first half of this suggested correspondence is underlined by the prediction of analysts of bureaucratic politics that, because of their size and unwieldiness, bureaucracies need to develop standard operational procedures in advance of engaging in action. This does not seem an unreasonable way of looking at the Single Integrated Operational Plan.

If there is nothing in the Ball/Snow approach that is exclusively nuclear, equally there is nothing in it that is exclusive to the United States (Snow is an American scholar, Ball an Australian), except for an ingredient in the atmosphere of political thinking in the United States that can incline analysts of the domestic political scene almost instinctively to look for instances of, or at any rate instances of resemblance to, 'separation of powers'. A notable and unusually explicit instance of this is the work of Don K. Price, a United States' analyst of science policy, who divided the scientific policy community in the United States into four 'estates' and saw a satisfactory relationship between science and society as conditional upon there being a balance between the estates.[11] Denying a specifically United States' application of the Ball/Snow approach is not the same thing as saying that it would not need some modification before being applied to noticeably less plural polities. It would be inappropriate, for instance, to apply it as it stood to the strategic nuclear policy of what was the Soviet Union in the Cold War period, although it would be wrong to condemn any such undertaking in advance as likely to be wholly unproductive.

The application of the Ball/Snow approach that is being suggested here involves a much more modest transposition of place and polity - from the United States to Britain - and a modest transposition of topic - from nuclear to conventional - specifically to the conventional defence policy of Britain in the period since 1970.

Before proceeding further, some tidying up is necessary. To begin with, it is almost but not quite 100 per cent clear what the Ball/Snow approach is saying. The preceding summary of their thinking shows that their vocabulary is similar but not identical. Where Ball speaks of three elements of policy (earlier, in 1981, he describes strategic nuclear policy as being divisible into facets, and, closely echoing Nitze, specifies the two most important as declaratory policy and action policy), Snow

11 Don K. Price, 'Money and Influence: The Links of Science to Public Policy', *Daedalus*, 103, 3, 1974. D.K. Price's four estates are practising scientists, administrators, professionals and politicians.

considers three levels. Where Ball speaks of one element having precedence - the employment or action element - Snow, in spite of speaking of levels, which is a word that would seem to imply an ordering into upper and lower, seems happy not to grant any precedence. In this book we shall use 'level' in the same neutral sense: this is not to deny the possibility that some ranking of levels may turn out to be justified.

Levels correspond to each of the alternative but not mutually exclusive ways in which an analyst, who we can assume for the moment is above the fray, can seek to understand policy. The other sense of level corresponds to the different ways of looking at the world of Britain's conventional defence according to which part of the strategic community the person in question belongs to - that of the professional military or that of the defence scientist, for instance. This second role for levels of policy should not come as a surprise. It corresponds to G. Allison's aphorism (which he claims he got from Don K. Price) summing up the tendency of bureaucracies to imbue their members with a particular way of looking at things: 'where you stand depends on where you sit'.[12] And its existence is a precondition for what in France is called 'professional deformation' - the conditioning of members of a professional group by the fact of their membership.

Unsurprisingly, perhaps, Snow, and more implicitly Ball, postulate a special place for the defence intellectual or academic in all this. According to Snow, the academic as a breed operates chiefly within the declaratory level. Shafer says very much the same sort of thing: 'academics may play three overlapping roles ... : the deep source of policy-makers' perspectives; legitimisers of (policy-makers') actions; and ... articulators of specific versions of widely believed notions'.[13]

Outsiders' perspectives

But at the same time, as the examples of Ball and Snow, among others, show, the occasional individual and presumably atypical academic is also capable of rising above his or her level and looking to comprehend the

---

[12] Graham T. Allison, *Essence of Decision: Explaining the Cuban Missile Crisis*, Little, Brown and Company, Boston, Mass., 1971, p. 176. Perhaps a fairer attribution would be 'anon'.

[13] Shafer, *op. cit.*, pp. 11, 12.

situation as a whole. This ability of an academic commentator to avoid being imprisoned within his normally appropriate level of policy may be triggered by some accident. The fact that Ball is not an American may have given him a freedom to see the American situation in a way that American-bred academics might have had to strive more consciously to attain. This outsider's advantage may also be part of the explanation of why American and Australian scholars have historically found it possible to produce excellent general studies of British defence policy - one can cite for instance Andrew J. Pierre's *Nuclear Politics*,[14] Philip Darby's *British Defence Policy East of Suez 1947-1968*,[15] or, from an earlier period, William P. Snyder's *The Politics of British Defence Policy 1945-1962*.[16]

A different but related point can also be made with respect to the advantages of perspective possessed by academic commentators who have moved or been transferred from one level of policy to another. Paul Nitze was out of United States' government service when he wrote his 1956 article, but had been George Kennan's successor as director of the Policy Planning Staff in the State Department from 1950 to 1952, and was of course later to return to government service. In Britain at the time of writing the leading academic commentator on national defence policy, Michael Howard, and his predecessor in broadly the same role, Basil Liddell Hart, both knew military service as professional army officers before exiting at a comparatively early stage in their military careers (both as it happens at the rank of Captain) from the operational level - that appropriate to an army officer - to take up the role of academic commentator (and military historian), for which work both received public marks of honour. Indeed were there not at least a handful of persons working in the academic world at any one time whose earlier career had included some acquaintance with policy levels other than the declaratory, the defence literature would be much less rich than it is and the screens of secrecy dividing the declaratory from the other two levels a good deal more opaque than they are.

---

14 Andrew J. Pierre, *Nuclear Politics: The British Experience with an Independent Strategic Force 1939-1970*, Oxford University Press, London, 1972.

15 Philip Darby, *British Defence Policy East of Suez 1947-1968*, Oxford University Press, London, 1973.

16 William P. Snyder, *The Politics of British Defence Policy, 1945-1962*, Ernest Benn Ltd, 1964.

Circles, levels and British defence policy

Introduction

Those who inhabit what we have described above as levels have a close resemblance to what Richard Little, in a theoretical contribution to a recent book on British foreign policy, calls a 'circle of social actors'.[17] Within these circles, the actors in question[18] have a particular way of conceiving an action or a state of affairs while remaining in a state of readiness to do things that will, they believe, directly or indirectly bring about the realisation of their particular goals. Little's name for this behaviour is 'practical theorising' though he considers that academic analysts can reasonably stand outside the circle, or in our case circles, and criticise these practical theories. There are powerful resonances here with the Ball/Snow idea of levels in the context of defence policy, although with a greater readiness to give the academic commentator a more privileged position than perhaps Ball or Snow would concede. The advantage of Little's circles is that they can be ascribed goals, which at a minimum must include, presumably, survival. In addition, coherence between the goals and the 'practical theorising' of one circle and another cannot be expected to be automatically forthcoming.

No attempt in the present study is made to build in any very systematic way on these theoretical positions. Rather, vocabulary, ideas and concepts will be freely borrowed from sources wherever they seem to add something worthwhile. In other words, few excursions will be attempted beyond the outer limits inherent in a straightforward adaptation of the original Ball/Snow analysis of US nuclear policy.

As far as British conventional defence policy is concerned, the three levels of policy put forward here are the declaratory, the operational (equivalent to the operational or employment or action levels of Ball and Snow), and the development-industrial - that is to say the whole business of producing weapons and their components from the research phase

---

17 Richard Little, 'Conclusion', in M. Smith, S. Smith and B. White (eds), *British Foreign Policy: Tradition, Change and Transformation*, Unwin Hyman, London, 1988.

18 Adapted from a dictionary definition of hermeneutics (a discipline social science has borrowed from biblical scholarship). Actually two definitions are used. See the entries under 'hermeneutics' and under 'intention' in *A Dictionary of Philosophy*, Pan Books, London, and Macmillan Press, London and Basingstoke, 1979.

through to their going into service. This last level is obviously based on the Ball/Snow notion of 'development' but has had its boundaries considerably extended to allow for the fact that we are now talking not of nuclear weapons but of conventional weapons. The difference between the two is not in the gross amount of scientific or engineering expertise that goes into the respective development of these weapons. This particular difference happens in fact not to be very great. In the case of Britain, official secrecy makes it difficult to give a precise figure for the breakdown of defence research and development expenditure (R & D) as between nuclear and conventional weapons, but the nuclear part of defence R & D is probably between 20 and 50 per cent of the total.[19] In other words the proportion of defence scientists and engineers working on conventional as opposed to nuclear projects could be anywhere between 4 to 1 and 1 to 1. The difference between the development-industrial level and the Ball/Snow 'development' lies in two other directions.

First, the production of conventional weapons has a much larger impact on the economy as a whole, for instance accounting for far more industrial employment than nuclear weapons. Again it is difficult to be precise, but for the period April 1989 to March 1990, expenditure on the British nuclear strategic force was, in current (1989/90) pounds, about £1158 million (not quite all of it equipment-related), whereas equipment-related expenditure for the armed services as a whole was £8536 million.[20] Assuming these proportions are accurately reflected in defence industrial employment, conventional forces account for about 86 per cent of the total. Bearing in mind that British nuclear expenditure was reaching a peak in the late 1980s and early 1990s (the proportion of defence industrial employment due to the production of conventional military equipment in 1985/86 had been as high as 94 per cent),[21] non-nuclear defence industrial employment outweighs nuclear, typically, by about 10 to 1.

Secondly, quite unlike nuclear weapons, conventional weapons are highly marketable internationally and, as such, normally have a part to play both in creating export earnings and in acting as an adjunct to more

---

[19] Philip Gummett, 'Problems for UK Military R & D', in Ian Bellany and Tim Huxley (eds), *New Conventional Weapons and Western Defence*, Frank Cass, London, 1987.

[20] *Statement on the Defence Estimates* 1991, Vol. 2, Cm 1559-II, London, HMSO.

[21] *ibid.*

general foreign policy objectives.

The declaratory level is very similar to the one identified by Ball and Snow. It is the officially stated description of what British defence policy is (we are nowhere in this book, apart from occasional brief excursions, directly concerned with British policy towards its nuclear weapons, and 'defence policy' will often be used as a shorthand for conventional defence policy). Declaratory policy is found, issuing from the horse's mouth so to say, in the annual statements published and put before parliament by the government of the day to accompany the defence estimates (budgetary) for the forthcoming year. The language of the declaratory level is also that in which government ministers couch their public speeches at home and abroad and statements to the press on defence matters. And it is also that in which defence debates in parliament are conducted. In other words opposition spokesmen operate within the same level. Also operating within the same policy level are some government advisers outside and within the civil service; but other government advisers, especially those within the civil service, will be found operating chiefly in one or more of the other two levels. If, for the moment, we follow Ball and Snow, we should also have to say that more wholly within the declaratory level are notionally independent commentators on defence writing in the press and elsewhere, and the bulk of the community of academic specialists on British defence and strategic matters. Most books on British defence policy, accordingly, are also written more or less within the declaratory level. Whether this statement may not need substantial qualification is a topic we return to below.

It is of course central to this kind of analysis that it will normally be more accurate, wherever one looks, to speak of strategic *communities* rather than a single strategic community. A simple visualisation, albeit inevitably rather static, can help to clarify some of the relationships between the three strategic communities that correspond to the three levels of British defence policy. Just as with Churchill's post-war description of Britain's foreign policy as lying in three interlocking circles (Europe, the Empire and the 'special relationship' with the United States), the three strategic communities corresponding to our policy levels also lie in three intersecting circles. The circles are now indeed 'circles of social actors'. More precisely, they are a development of the same idea. They can still be distinguished by their inhabitants, corresponding to three strategic communities, except that some inhabitants occupy two circles rather than one, and a few can simultaneously occupy all three.

The population of the circles

The declaratory circle we have already peopled. The operational circle also includes some government politicians and their advisers, but chiefly consists of the professional military. It too is not very different from the Ball/Snow description except that it is more densely populated. That is to say, it contains a higher proportion of the professional military establishment than would be the case if we were speaking here of the circle corresponding to the strategic nuclear level, which in all nuclear powers is inhabited by only a small fraction of the total military establishment, at least at any one time.

The development-industrial circle also resembles its Ball/Snow equivalent, but in this instance our transition from 'nuclear' to 'conventional' is even more significant. This circle certainly contains some government politicians and their advisers, together with some military who may be retired rather than active, scientists and engineers working for government defence organisations or within industry, or even in academic life, but it also very importantly includes industrialists and trade union leaders connected to firms in the defence sector of industry. Here Snow's simple shared dominant roles within his nuclear development circle of scientists and engineers in the design of weapons and of politicians in the procurement of the weapons have to be reconsidered in the light of the much larger role the production of conventional weapons plays in the industrial sector of the economy.

The manufacture of conventional weapons in Britain is big business. During the period 1980 to 1990 it was responsible, year in year out, on average for the direct employment of about one-quarter of a million persons (using data published by the Ministry of Defence, and including employment arising from exports of armaments; this estimate does not include employment on nuclear weapon projects).[22]

---

22 Average direct employment in Britain arising from the production of military equipment purchased by the MOD over the ten-year period was 200,000. Subtracting the proportion of this due to nuclear work - about 10 per cent - we are left with 180,000. Direct employment arising from exports of military equipment has to be added to this figure. This brings the total to 250,000. *Statement on the Defence Estimates*, Vol. 2, Cm 1559-II, 1991, HMSO, London.

The significance of the defence-contracting industry in Britain is not just a matter of quantity. A 1985 study of British industry of more than usual objectivity judged the following firms as being able to match international standards of competitiveness (excluding oil companies) - to be, in other words, 'of global stature' - Pilkington, BOC, Reuters, Beecham, Racal, Plessey, Glaxo, British Aerospace, Rolls Royce and 'parts of' GEC.[23] Whilst according to a commercially confidential study carried out in 1987 for the Ministry of Defence and the Department of Trade and Industry by General Technology Systems Ltd, of these ten, five - British Aerospace, GEC, Rolls Royce, Plessey and Racal - are, in that order of turnover, the five chief producers of military equipment in the country. A sixth company of global stature, Pilkington, was just outside the top ten producers of military equipment.[24]

The declaratory level

At first sight it may be thought that policy made within the declaratory level - declaratory policy - can be altered suddenly, in a way not possible either for development policy, as Snow points out, considering the very long lead times characteristic of modern weapons and their components (whether nuclear or conventional) or for operational policy, which is to a large degree constrained by the weapons available. Declaratory policy is, after all, only a form of words. In practice this seems to be both true and untrue. It is true in the sense that no two statements on the defence estimates published by the British government are exactly alike. On the other hand, even over a twenty year period, the same phrases and sentiments show a marked tendency to surface time after time. It is as if declaratory policy were a carefully wrought multi-role formula for sending appropriate signals to real and potential enemies abroad, as well as for reconciling conflicting elements within the ruling party and within the country; and while reformulations are permissible within limits, the basic formula should therefore be

---

[23] John M. Stopford and Louis Turner, *Britain and the Multinationals*, John Wiley and Sons, Chichester, 1985, p. 105.

[24] ACOST (Advisory Council on Science and Technology, Cabinet Office) *Defence R&D: A National Resource*, HMSO, London, 1989, p. 55.

fundamentally tampered with and these limits overstepped only at one's peril. In addition, declaratory policy is subject to two very important constraints, the outward and the inward.

The outward requires that the declaratory defence policy should be in harmony with the stated foreign and economic policies of the government. Moreover, where allies are important - and for a medium power like Britain they are - there is the additional outward constraint that requires declaratory policy to be in step with the defence policies of allies.

Concerning the inward constraint, we have had enough clues from the Ball/Snow approach to United States' nuclear policy to know what elements are at work. The declaratory level must be consistent and in some sort of harmony with the other two levels, in that the armed forces that are or can soon be made available place bounds on the freedom to reshape declaratory strategy at will. There are three related factors at work determining the rate at which available forces can be changed. A long operational lifetime of modern weapons is usually judged necessary to defray the very high cost of procuring them in the first place. This induces an obvious static quality. Attempts to acquire new weapons can be made, but will be slow to take effect: manufacturing new weapons from research and development through to entry into service takes a long time - ten years is not unusual - indeed this is one reason why they are so costly. And even an attempt to get rid of weapons and to dispense with any requirement for more cannot itself be executed other than slowly, if an orderly domestic transition from a defence policy which implied the direct and indirect employment of a very large number of persons in the civil economy to one that does not is desired (in Britain, indirect employment adds about 80 per cent to the figure of a quarter of a million already given for direct industrial employment, including persons employed in the manufacture of armaments for export)[25]. Within these rather inelastic bounds, policed so to speak by the population of the respective operational and defence-industrial circles, it seems reasonable to suggest that while small changes to declaratory policy can be made reasonably quickly, large changes are

---

[25] This is a typical figure for the period 1980 to 1990. See *Statement on the Defence Estimates*, 1991, Vol. 2, Cm 1559-II, HMSO, London, p. 61. See also Ian Bellany, 'UK Defence Expenditure: Its Impact on Jobs and Energy Use', *Bailrigg Paper* No. 8, Lancaster, 1985.

much more difficult to implement in the short term. Lest this should be thought of as ascribing a subsidiary rank to declaratory policy, consistency between it and the other two levels of policy is a two-way street. Just as declaratory policy cannot get too far ahead or fall too far behind the other two, neither can they fall very far out of step with the declaratory level.

The defence review

But great reformulations of policy in Britain do occur. They are called 'defence reviews' and, as one would expect from the foregoing, are quite rare (all great reformulations of policy are called defence reviews, but not all defence reviews are great reformulations of policy). The two most recent defence reviews at the time of writing were nine years apart, 'Options for Change' in July 1990 and 'The Way Forward' in June 1981. Interestingly, the latter was radical in intention but was overtaken by the war in the Falklands and never took proper effect; the former may have aspired to match the radical ambitions of 'The Way Forward', but it was not universally regarded as having done so, the government being, in the words of one academic commentator, 'determined not to disrupt the essential features of the nation's defence roles'.[26]

The defence review, in fact, is central to an understanding of the interaction between our three levels of defence policy. Defence reviews (of the radical sort) become necessary when the three levels - the declaratory, the operational and the development-industrial - get too far out of alignment, or, in other words, where the elements of contradiction as between the three levels, and within them, become excessive. The defence review is a process whereby an attempt is made to bring the three levels of policy into line. This may lead to the publication of a new declaratory policy, but need not necessarily do so.

Nuclear digression

An interesting illustration - which is the one sizeable diversion we allow ourselves into nuclear politics - concerns the British strategic nuclear

---

[26] Paul Laurent, 'The Costs of Defence', in Stuart Croft (ed.), *British Security Policy*, Harper Collins, London, 1991, p. 101.

declaratory policy after the decision was taken to proceed with ordering a replacement for the Polaris flotilla. The initial announcement, to acquire the Trident C4 missile from the United States, was made by the Thatcher government in 1980, but this was modified in 1982 when the more modern, and expensive, Trident D5 missile was ordered instead. In the same year the opposition Labour Party conference adopted by a two-thirds majority a motion in favour of unilateral nuclear disarmament by Britain. This (as it transpired) temporary breakdown of bipartisan agreement on whether there should be a British strategic nuclear force at all was accompanied by a resurgence of domestic anti-nuclear protest. There could be no surer sign that declaratory policy was under strain and no coincidence that this strain should show at a time when changes were afoot in the operational and development-industrial levels. The last time the Labour Party conference had (again temporarily) broken the bipartisan consensus was in 1960, which was the last time the nuclear operational and development-industrial levels had undergone significant change. This occurred after the cancellation by the United States government of Skybolt - an American-designed ballistic missile for launching from heavy aircraft which Britain had been relying on to give its V-bombers a new lease of life.[27]

In the early 1980s, although not wholly created by it, popular protest gathered momentum with the 1979 NATO decision to station United States' intermediate-range nuclear missile forces (INF) in Europe, with Britain hosting ground-launched cruise missiles.

A reading of successive statements on the defence estimates over this period shows a slight but definite steering of nuclear declaratory policy away from presenting the British strategic nuclear force as essentially a contribution to NATO (even if sometimes proudly described as 'unique') towards the adoption of a more Gaullist guise, as a national force, to be followed by a shift back again as circumstances changed once more. Thus the 1985 statement on the defence estimates speaks of the nuclear force as 'assigned to NATO ... however our strategic nuclear deterrent remains at all times under the independent control of the British government'[28]. The 1987 statement shows interesting signs of movement in declaratory policy when it speaks of the 'independent

---

[27] Pierre, *op.cit.*, pp. 181, 182.

[28] *Statement on the Defence Estimates*, 1985, Cmnd 9430-I, HMSO, London.

nuclear deterrent under *absolute* British control' (emphasis added).[29] This was the high point of neo-Gaullism, for the 1991 statement is found once again to state that 'our independent nuclear force will be assigned to NATO while remaining at all times under the control of the British government'.[30] Of course in the meanwhile, at the operational and development-industrial levels, nothing had changed at all, except for the completion of a long drawn out semi-secret Chevaline programme to equip the Polaris A3 missiles with new British designed re-entry vehicles. But the point is that real imminent change at these levels had been signalled.

The intricacy of the web of constraints on declaratory nuclear policy can be nicely illustrated starting with a point due to Stuart Croft. Shifting the emphasis of the planned Trident flotilla from 'national' to 'NATO' and back, according to Croft,[31] was determined by a desire to keep the British force out of the reach of whatever kind of nuclear arms control negotiation seemed highest on the international agenda at the time (off the agenda altogether since 1979, pretty well, nuclear arms control returned to the realm of practical politics in 1985). If the topic of the moment was theatre nuclear forces, then Polaris and its Trident successor were 'national', but if strategic nuclear arms control threatened to come to the fore, then the British force became 'NATO'.

But we know this is not the whole story. In a period of substantial public protest against nuclear weapons, the national nuclear force, with the emphasis on the word 'national', was always backed by a majority of public opinion. In early 1982, according to opinion polls, the British national nuclear force was approved of by about 58 per cent of the public: a year later the figure had risen to 65 per cent. By contrast, a poll taken during the same period put *opposition* to the stationing of United States' land-based cruise missiles in Britain at 54 per cent.[32] So government was tempted to shape declaratory strategy accordingly. On the other hand, the reason for the popularity of the SLBMs as a national force may have been an expression of the anti-Americanism that

---

29 *Statement on the Defence Estimates*, 1987, Cm 101-I, HMSO, London.

30 *Statement on the Defence Estimates*, 1991, Cm 1559-I, HMSO, London.

31 Quoted by Hoffman, 'From Conformity to Confrontation: Arms Control', in Croft *op. cit.*, p. 70.

32 David Capitanchik and Richard C. Eichenberg, *Defence and Public Opinion*, Royal Institute for International Affairs, Routledge and Kegan Paul, London, 1983, pp. 20, 24.

presumably contributed something to the unpopularity of the NATO-dedicated United States' ground-launched cruise missiles the government had agreed to accept. If so, the 'national' flag was not something that could be waved too vigorously within the declaratory strategy, when the operational and development aspects of the Polaris force and even more so its Trident replacement depended so much on continued United States' cooperation. Incidentally, lest it might be thought that this sort of fine-tuning of nuclear declaratory policy would in practice be difficult for government to manage, it is the case that the same cabinet committee, with presumably the same *ex officio* membership, MISC 7, was responsible for handling both the Trident and INF issues.[33]

But, as we shall see below, the limits within which declaratory strategy is free to alter are not always so narrow.

Conventional declaratory strategy

Introduction

Overall the declaratory British conventional defence policy for the past twenty years has perhaps varied even less than the nuclear. Most of the variation has been in the degree of emphasis given to what is nonetheless consistently presented as the essentially subordinate and secondary question of defence policy outside the NATO area. New Conservative governments have usually begun by giving somewhat more emphasis to this than their Labour predecessors, before settling back to somewhere slightly above the mean level. New Labour governments have usually started by giving this less emphasis than their Conservative predecessors, before they too return to the mean or a little below it.

Thus the 1977 statement on the defence estimates ran as follows:

> The defence of the United Kingdom remains firmly based on the North Atlantic Alliance ... The Government's policy is to concentrate its defence effort on the Alliance and to deploy the Armed Forces of the United Kingdom in areas in which they can make the most significant contribution to the security of the

[33] Michael Clarke, 'The Soviet Union and Eastern Europe', in Peter Byrd (ed.) *British Foreign Policy Under Thatcher*, Philip Allan / St Martin's Press, Oxford, 1988.

alliance and, through that, to national security; it judges these to be the East Atlantic and the Channel, the Central Region on the mainland of Europe, the security of the UK base, and the strategic nuclear deterrent.[34]

The 1977 statement did not find it possible to say anything at all about the secondary, extra-NATO ('out-of-area') role for British armed forces, but by 1979 something like a middle position had been reached, with a reference in the statement of that year to the country's 'ability to deploy its maritime forces world-wide in defence of Allied or national interests ...'[35]

Flushed with the success of the Falklands campaign, the Conservative administration's 1983 statement overshot the middle point by some way. After recognising the primacy of NATO in the country's defence policy and reaffirming the four main roles the country plays in NATO - precisely the four referred to six years earlier, in the 1977 statement - the 1983 statement goes on to announce:

> we are planning to make a number of improvements in the capability of our NATO forces (*sic*) to operate outside, as well as in, Europe. We are also taking further steps to improve the coordination and effectiveness of these diverse activities which, although subordinate to our primary NATO roles, form an important component of our overall defence policy.[36]

But reversion to somewhere near the middle point was on its way. According to the 1988 statement on the defence estimates:

> British defence policy remains founded on membership of NATO ... Over 95% of the defence budget goes, directly or indirectly, towards carrying out our main defence roles in NATO. The great majority of our forces are committed to one or other of the three main NATO commands, Europe, Atlantic and Channel; and we remain the only European member of the Alliance to contribute to

---

[34] *Statement on the Defence Estimates*, February 1977, Cmnd 6735, HMSO, London.

[35] *Statement on the Defence Estimates*, February 1979, Cmnd 7474, HMSO, London, p. 18.

[36] *Statement on the Defence Estimates*, 1983, Cmnd 8951-I, HMSO, London, p. 3.

> all three elements of NATO's triad of forces: strategic nuclear, theatre nuclear and conventional. We also have important defence commitments outside the NATO area ... They include a direct responsibility for the defence of our remaining dependent territories. They also include activities ... which help to achieve the Government's foreign policy aim of promoting stability in areas where we have important economic interests. And since our interests coincide closely with those of our allies, these efforts make an important contribution to preserving broader Western interests outside the NATO area.[37]

And back again, to 1973:

> Although Britain's basic security depends on the strength of the North Atlantic Alliance and it is to NATO that by far the greater part of Britain's military forces is committed, Britain continues to play her part, within her resources, in countering threats to stability outside Europe. For without that stability Britain's world-wide political and trading interests cannot flourish.[38]

But it would be wrong to think nothing changes. Perhaps had the Falklands War not intervened, the Nott defence review of 1981 ('The Way Forward') would have been implemented,[39] and the corresponding changes of naval strategy at the operational level - with more submarines and greater reliance on land-based aircraft - would have had significant consequences for the navy and for its suppliers at the development-industrial level. This, as we shall see, would have had few declaratory consequences since it would have been a question of adjusting the operational level to bring it in line with the declaratory. Even so, declaratory policy outside the main NATO element has changed and sometimes with remarkable speed.

The 1987 statement on the defence estimates seems to mark a remarkably sharp shift in declaratory policy[40] when it claims 'none of

---

37 *Statement on the Defence Estimates* 1988, Cm 344-I, HMSO, London.

38 *Statement on the Defence Estimates*, February 1973, Cmnd 5231, HMSO, London.

39 'The Way Forward', Cmnd 8288, HMSO, London, 1981.

40 *Statement on the Defence Estimates*, 1987, Cm 101-I, HMSO, London.

our out-of-area garrisons is constituted with Soviet aggression in mind or even aggression by proxies of the Soviet Union'. This is a substantial change of tune from what was being said as recently as 1985, and its suddenness caught out at least one commentator.[41]

Costs

The declaratory level of British defence contains something else. There is a broadly shared assumption within the declaratory strategic community that the expenditure of public money on defence should be kept under close scrutiny since the defence budget is a burden upon the back of the country. Thus, in 1977:

> The Government has ... found it necessary in the past year to reduce planned defence expenditure as part of its policy to reduce public expenditure generally. To bring down the level of public sector borrowing and to free resources for industry, the Government announced ... a reduction ... in public expenditure planned for 1977/78.

And in 1980:

> The Government is committed to making the best possible use of the taxpayer's money. The Ministry of Defence ... must play a full part in the pursuit of efficiency and savings ... The resources available for defence are limited and we cannot afford to waste them.

Or, in 1988:

> (It is) important to pursue efficiency and value for money to the greatest possible extent. The Ministry of Defence has made a major contribution to the Government's plans to reduce the size of the Civil Service ... The search for further economies will continue. Economies are also being made in Service manpower.

The view within the declaratory circle is that expenditure on defence

---

[41] C. Bluth, in Croft, *op. cit.*, p. 62.

plus expenditure on everything else can only add up to a fixed, or at best, a slowly increasing sum. The more that defence takes up, the less there can be for everything else.

This brings the declaratory level and its circle of social actors up against the development-industrial level and its circle and calls for reciprocal adjustments. We learn from Peter Byrd that since 1979 the government has attempted:

> to introduce greater discipline and market forces into the management of defence in general and defence procurement in particular ... The government placed more contracts by competitive tender and contracted out (privatised) the provision of some services. It is difficult to evaluate the success of this policy, though there was a major shift from 'cost-plus' contracts towards competitive tendering ... the Select Committee on Defence has remained sceptical about the success of the policy.[42]

We would not be taking a great risk if we inferred that resistance to the policy on the part of the defence-contracting industries was one reason for the success of the policy not being immediately apparent to the informed observers of the Select Committee.

We would not expect jostling between the declaratory and development-industrial circles to be limited to purchasing methods. About a quarter of all scientific research and development in Britain is performed by or in relation to the development-industrial circle (see the Appendix to Section Three). This is sometimes seen as disproportionately high when the defence budget itself is typically between 4 and 5 per cent of the Gross Domestic Product (GDP). The widely supposed importance of national research and development to the economic health of the country leads to tension between the declaratory and development-industrial circles. The pure declaratory position is reflected in the 1987 statement on the defence estimates where we read that 'the government shares the underlying concern of those who fear that necessary investment in defence R & D may "crowd out" valuable investment in the private sector'.[43] The digging in by the development-

---

[42] Peter Byrd 'Defence Policy' in Peter Byrd (ed.), *British Foreign Policy under Thatcher*, Philip Allen/St Martin's Press, Oxford, 1988, pp. 174-175.

[43] *Statement on the Defence Estimates*, 1987, Cm 101-I, HMSO, London.

industrial circle is however reflected in the compromise announced in 1985. In that year the government set up a body called Defence Technology Enterprises in an attempt to improve 'the spin-off to the civilian market place of innovative ideas and technology that arise in the defence research programme', saying that 'we are looking at the relative priorities of civil and defence needs for scarce scientific and technical resources.'[44]

Circles within circles?

It seems appropriate at this point to emphasise something that should have become apparent. Claiming that there is a declaratory level of policy and a reasonably identifiable strategic community working and thinking within it is not at all the same thing as saying that the members of this community are all of the same mind. Returning for a moment to the Ball/Snow analysis of the United States' strategic nuclear policy, there have always been arguments about that policy within and at the declaratory level. Most famously perhaps, when declaratory strategy was 'assured destruction and damage limitation' there was a flurry of argument over the meaning of 'damage limitation'. Within the declaratory level of British defence policy there are periodic arguments about the balance between nuclear and conventional. There are arguments about where the internal weight within conventional defence policy should be distributed as between the principal declaratory role of being an effective member of NATO and the secondary declaratory role of defending interests outside the NATO area, and about what ranking to accord these interests. It is these arguments, and arguments like them, that make writing and commentary done from within the declaratory circle worthwhile and worth reading. Whether these arguments are conducted within parties or across parties, academic and journalistic commentators as well as publicists and pressure groups are also more than willing to become involved.

In much the same way, the other levels of defence policy, the operational and the development-industrial, are also characterised by internal disputes and arguments. The difficulty is that the screens of secrecy behind which members of these levels operate ensure that only a little becomes publicly visible. On the other hand safe inferences are

[44] *Statement on the Defence Estimates*, 1986, Cmnd 9763-I, HMSO, London, p. 50.

often possible.

Whatever view may ultimately be taken of the defence-contracting industry in Britain, there is at least some commercial competition and hence rivalry between firms for MOD and overseas business. There are, naturally, disputes between management and labour. And proposals by larger firms for the takeover of smaller ones are not always welcomed by the latter with outstretched arms. Other transactions between firms of more comparable weight do not always proceed smoothly either. In 1990 and 1991 Ferranti found itself in a dispute with GEC over the sum paid by the latter for its purchase from the former of Ferranti Defence Systems.[45]

Disagreement within the operational circle also occurs. The expression 'inter-service rivalry' was coined to describe it. Again, not much is conducted in public. Perhaps the most publicised instance within (only just) our period of enquiry concerned the defence review of 1966, where the view of the RAF that the future presence of British armed forces East of Suez would be better founded on their 'island plan' for a number of bases for long-range strike aircraft (by this stage seventy-five F-111 aircraft to be purchased from the United States), prevailed over the Royal Navy's plan for a presence based on what would probably have amounted to three British-built aircraft-carriers. Very quickly, by 1971 in fact, it turned out that the RAF victory was a hollow one, when the order for the F-111 aircraft was cancelled.[46] Equally, although it took much longer to become clear, the navy had simply lost a skirmish and eventually got their carriers, or something very like them, after all. Evidence that this quarrel went deep comes from the fact that after-shocks were still being felt as late as 1984, when a retired RN officer (Captain F.D. Stanley) proposed that the RAF Tornado aircraft should be sold to pay for a more active RN presence East of Suez.[47] Quarrels are not restricted to the regally-crested branches of the armed services. The RN, in the person of Admiral of the Fleet Lord Hill Norton, launched an attack on the army in 1983 with the suggestion that a more global role for Britain (centred on the navy) could be paid for by cutting

---

45 See *The Independent*, 12 December 1991.

46 John E. Woods, 'The Royal Navy since World War II', *Proceedings US Navy Institute*, March 1982.

47 Captain F. D. Stanley, 'The Royal Navy and its Future', *Navy International*, July 1984, p. 327.

back on the army's presence (and that of the RAF) in Germany.[48]

Indeed, not much imagination is necessary to envisage the strong possibility that quarrels are likely even within the different branches of the same armed service - for instance between the sub-surface and the surface arms of the navy.

Accordingly, it cannot be denied that there is something arbitrary in settling for three levels of policy; if each level (or circle of actors) is capable of further subdivision, why stop at three, it might be asked.

There are two kinds of answer to this. The first is to ask for a suspension of judgement. If the reader finds that in the end he has learned something about the subject from the 'three levels' treatment given here, that is one kind of justification for its adoption in the first place. One can imagine this sort of justification being put forward on behalf of the Ball/Snow characterisation of United States' nuclear policy in the event of its being pointed out (quite correctly as it happens) that the United States' military - particularly the navy and the air force - has a long record of quarrelling between themselves on nuclear matters. This does not of course rule out the possibility that subsequent work could increase the number of levels, or circles, and that greater complexity could be rewarded by greater realism.

The other answer is to suggest parallels with the techniques employed with such success by natural scientists. Indeed the framework for analysis adopted here will sound to some ears very like an attempt to follow the methods of the earliest of the modern natural scientists (as we should now call them) who seem to have believed in the 'way of resolution':

> The intuitive idea which informs this methodological tradition was this: the way to understand something is to take it apart, in deed or in thought, ascertain the nature of its parts, and then reassemble it - resolve it and recompose it.[49]

Thus, for instance, simply because matter is capable of very extensive, if not limitless, subdivision into ever more microscopic 'parts', this does

---

[48] Admiral of the Fleet Lord Hill Norton, 'Return to a National Strategy', in John Baylis (ed.), *Alternative Approaches to British Defence Policy*, Macmillan, London, 1983.

[49] J.W.N. Watkins, *Hobbes's System of Ideas*, Hutchinson University Library, London, 1964, p. 52.

not preclude interesting and internally consistent observations concerning its properties by focusing alternately on matter at different levels of aggregation. The possibility of sub-atomic physics does not preclude the possibility of chemistry.

Beyond the declaratory level

Nonetheless, the obvious fact that so much vital disagreement, debate and analysis do occur within the declaratory circle of defence policy can tempt an observer to think that the declaratory is virtually all that there is and nothing much else matters. This would be a mistake for two reasons.

The first, echoing Snow on United States' strategic nuclear policy, involves the recognition that the declaratory policy level is by definition public and open whereas much thinking about and discussion of military affairs must, as we have already hinted several times, clearly take place in private. At both the operational and development-industrial levels this must be so for national security reasons. Information concerning the performance of weapons and details concerning the employment of force will be released into the 'public domain' only on a highly selective basis so as to maximise the desired impact on potential enemies. This may involve not merely the withholding of genuine information but in addition the dissemination of untruthful and inaccurate statements, the better to mislead opponents. Absence of genuine information plus the presence of disinformation together make quite an effective screen of secrecy. It has to be admitted that the rare public manifestations of inter-service rivalry is in no way indicative of the rarity of the phenomenon itself, since a public image of unanimity clearly avoids giving comfort to any potential enemy. The development-industrial circle also has its own reasons for conducting its affairs in secret. Commercial secrecy may be crucial to the profitability of a defence-contracting firm, whether it wishes to hide an innovatory product from a competitor at home or overseas, or to disguise its true level of profit from a customer or a potential customer, or even from a sister company with which it may be arranging transactions of some kind.

None of this is to say or even imply for a moment that the 'real' strategy is the secret or more secretive one(s), and that the public declaratory strategy is part of a mysterious but organised and deliberate attempt to mislead the hapless citizen. Much of British defence policy *can* be understood within the declaratory level, but not everything. The same qualified claim could be made in principle for the other two levels. In practice the present author knows only of instances of attempts to understand defence policy from the joint perspective of the two non-declaratory levels rather than from one or the other separately. The result is then none other than an analysis of defence policy from the perspective of the 'military industrial complex'. An attractively written and combative study of British defence policy which comes close to fitting this description is Malcolm Chalmers's 1985 book *Paying for Defence*.[50]

As stated in the Introduction, it is the puzzle-solving power of going beyond the declaratory level that is our ultimate justification for embarking on the journey. But before elaborating on this it is useful to take a detour in an attempt to complete a little unfinished business - to solve a small puzzle raised in the present section. Is the British academic/student of military and defence affairs going to be able to free himself from the declaratory circle only in rare and unusual cases, as Snow would seem to suggest, or is Little more accurate in according the academic a privileged role as a disinterested observer?

Academic awareness reconsidered

First, a summary of declaratory policy which has done service virtually unchanged between the early 1970s and the late 1980s can be derived from the above. Since defence spending is a burden, the country can afford less of it than once was the case. Value for money in the defence budget is the important criterion. Collective defence in Europe alongside the NATO allies is the chief priority, with as much as 95 per cent of British military effort directed there. A small residual capacity is retained for warlike purposes in parts of the world not within the geographical and strategic remit of NATO, but even 'out of area' *national* goals and objectives have to take their place alongside (sometimes slightly ahead of, sometimes slightly behind) the contributions British forces make to *collective* purposes 'out of area'.

---

[50] Malcolm Chalmers, *Paying for Defence: Military Spending and British Decline*, Pluto Press, London, 1985.

British academic specialists

How far has the British community of academic specialists in defence been able to break free from their supposed membership of the declaratory circle and look beyond it? Does the evidence support the Snow thesis that it is an unusual academic who can do this, or is Little correct in his belief that the academic is a privileged observer normally capable of standing outside and commenting objectively on our three intersecting circles of social actors?

Whilst it is true that there is no previous academic study known to the present author which even begins to make explicit distinctions between declaratory and operational levels for British defence policy, Snow's position has an inevitably self-negating quality. Once his ideas along with those of Ball and of other kindred analysts had entered the academic bloodstream, it was almost inevitable that academics would thenceforth start to look at the world a little differently. Little himself is a case in point.

Another example, admittedly indirectly suggestive of the correctness of what has just been said, comes from Michael Smith, writing in 1988 on British foreign policy. With reference to the British attitude to the demands made by the United States that it should follow the anti-Soviet line of the Carter administration over Afghanistan and later, in 1981, the Reagan policy of imposing sanctions on the Soviet Union in retaliation for the Polish declaration of martial law, Smith sees Britain as seeming to side with the United States, but in practice following the European Community. As he puts it, 'declaratory and operational [foreign] policies followed different paths'.[51]

With regard to British defence policy as such, Mick Dillon, also writing in 1988,[52] is far from content with a simple rational actor model of policy-making. He speaks of there being within the Ministry of Defence 'talk' of different kinds - 'services talk, procurement talk'- and he refers to 'ministers mediating between the public and private faces of decisions'. The former would seem to dovetail with our reference to documentation appropriate to the different levels of defence

---

[51] Michael Smith, 'Britain and the United States: Beyond the "Special Relationship"?', in Byrd *op. cit.*

[52] 'Britain', in G. M. Dillon (ed.), *Defence Policy Making: A Comparative Analysis*, Leicester UP, 1988.

policy, and the latter certainly sounds like what we have described as bringing levels of policy into alignment and the unique role played by senior governmental politicians in this matter. On the other hand, Dillon does not really seem to be suggesting the coexistence of different levels of policy and their circles of actors. Instead he would seem to be extending in an exploratory way the Allison bureaucratic model of decision-making with the Freedman gloss - that it is at the tactical level, below the topmost strategic level of decision-making, that all the bureaucratic pulling and hauling takes place.

Resources and 'commitments'

Other writers on British defence policy do not come quite as close as Dillon to anticipating what is said here about strategic communities and circles of social actors. But a number have always seen the central problem of British defence policy as a matter of 'matching commitments and resources', to quote from only one of the most recent commentators to employ the image, Paul Laurent.[53] Perhaps the most thoroughgoing treatment of British defence policy from this perspective is that due to Martin Edmonds, and his chapter entitled 'Planning Britain's Defence', in his *The Defence Equation*.[54] The title of the book is itself a reference to Edmonds's thesis that the defence equation to be balanced is that involving resources on one side and commitments on the other. The lineage of the image goes back at least as far as Walter Lippmann and his assertion fifty years ago that United States' foreign policy should be about bringing into balance 'the nation's commitments and the nation's power'.[55]

There is certainly an overlap between what Edmonds and others are saying and what is being said here about the necessity to bring into line the three levels of defence policy. In one way, if read literally, Edmonds is saying much less. Declaratory policy may *include* reference to what are described as commitments, to the defence of allies for instance, but even these arrangements do not always in reality (although it may from time to time suit the book of the operational circle to have

---

53 'The Costs of Defence' in Croft, *op. cit.*

54 Martin Edmonds (ed.), *The Defence Equation*, Brassey's Defence Publishers, London, 1986.

55 W. Lippmann, *US Foreign Policy*, Hamish Hamilton, London, 1943, p. 5.

it suggested otherwise) have the binding or obligatory quality transcending day-to-day calculations of interest that the word 'commitment' seems to imply. In another way Edmonds and others are saying more, even when we substitute 'power' with 'resources', for they would seem to be referring to the totality of what the state and all of its resources are 'committed' to externally (rather than its declaratory defence policy and such of its resources as may be relevant). Even so, by equating 'commitments' more or less to what is said in the declaratory policy statements about the objectives of defence policy and 'resources' to that with which the operational and development-industrial levels are in their different but related ways concerned, the matching of the two amounts to an aspect of keeping the three levels of policy in line. The Edmonds image also concedes, by implication, that such matching needs some sort of intervention and does not happen automatically - indeed what does happen automatically is the drifting out of alignment. Laurent's treatment of the aborted Nott defence review of June 1981 as an attempt at a radical rebalancing of commitments and resources removes any doubt as to the correspondence between his view and that taken in the present work as to the essential nature of the defence review.[56]

Other authors show a keen awareness that the drifting out of line or emergence of contradictions is not accidental. Writing in 1988, Tony McGrew gives full weight to the part likely to be played by the armed services in blocking the implementation of defence cuts. He also speaks of the defence contractors having:

> 'insider' status within the [Ministry of Defence], and naturally inclined to define defence needs in ways that complement their own commercial prospects.[57]

David Sanders, writing in 1990,[58] explains the puzzle of high British defence budgets (the high fraction of GDP spent on defence by Britain compared to that spent by its major European allies) by referring to the success of service and defence industry lobbies. He is doubtful about

---

56 Laurent, *op. cit.*, pp. 91, 95.

57 'Military Security and Order', in M. Smith, S. Smith and B. White, *British Foreign Policy: Tradition, Change and Transformation*, Unwin Hyman, London, 1988.

58 *Losing an Empire, Finding a Role*, Macmillan, London, 1990.

how much weight should be attached to the declaratory policy (in our terminology) of withdrawal from East of Suez, promulgated at the end of the 1960s and at the start of the 1970s, suspecting that had declaratory and operational policies (again using our terminology) been, or remained, in line, smaller defence budgets would have been the subsequent order of the day.

Oddly, neither McGrew nor Malcolm Chalmers, in the work we have already referred to, seems to have the same doubts about the declaration of withdrawal from East of Suez. The particular interest of Chalmers's work is that (in our terminology) it virtually amounts to a statement of the operational and development-industrial levels of policy, written around the supposition that these policy levels are in harmony but out of line (in our terminology) with the declaratory; moreover, that it is the former two levels that are in some important sense dominant, and constitute 'reality', whilst the declaratory merely conveys 'appearances'. He even prints a summary of 'real' British defence policy, at whose head is the maintenance of Britain's imperial role.[59]

What might be called a mirror image of Chalmers's work is provided by John Baylis's 1989 book, *British Defence Policy: Striking the Right Balance*. Where Chalmers ascribes predominance to the operational and development-industrial levels, Baylis (again using our terminology) gives the same predominance to the declaratory level. He demonstrates what has been contended already - that even if the declaratory level of defence policy is taken at face value, interesting academic studies of British defence are still possible.

Where not only Baylis but also Chalmers and virtually every other academic commentator cleave closely to the documentation of the declaratory level is in the assumption, supported by what Chalmers at one point seems to be conceding is very slight evidence,[60] that defence expenditure is a burden. Other commentators have no doubt whatever; Roger Tooze, in what may be an extreme example of the tendency, claims that it has been 'comprehensively demonstrated' that civilian industry in Britain has been adversely affected by heavy defence commitments.[61]

---

[59] Chalmers, *op. cit.*, p. 15.

[60] Chalmers *op. cit.*, p. 4.

[61] 'Security and Order: the Economic Dimension', in Smith *et al*, *op. cit.*

Greenwoodery and the conventional wisdom

What is often seen as the other important facet of the economic side of defence (although it is more than a matter of economics alone) does not command the same consensus. Some academics see the scale of Britain's military capability to have been undergoing a long, slow decline since 1945, which conforms to the view adopted within the declaratory circle. Others, probably a minority, resist this interpretation, finding more evidence of restructuring than of decline, and in at least one instance no evidence of decline at all. David Greenwood is probably the chief 'non-declinist',[62] along with Roy Jones.[63]

Greenwood is a former civil servant in the Ministry of Defence who has tended in his academic career to specialise in economic aspects of British defence policy. Writing in the 1970s and early 1980s, Greenwood argued that it would be hard to prove that Britain 'disposed of less military might' then than in the 1950s, and that by the late 1970s Britain was 'making provision for defence on almost exactly the same scale as in 1965 or 1955'.[64] Whilst probably in some sort of minority, Greenwood has allies. Writing in 1982, Woods finds that in the preceding decade 'the Royal Navy has been remarkably successful in maintaining its position both in strength of active units and balance between various types of combatants'.[65] And in much the same vein Wyllie suggests that in 1982 'Britain disposed of relatively more firepower than at any time since the Second World War'.[66]

The debate between the 'conventional wisdom' of decline versus the 'revisionism' of restructuring became a minor academic cottage industry in the UK for about ten or twelve years between the mid-1970s and mid-

---

[62] In John Baylis (ed.), *British Defence Policy in a Changing World*, Croom Helm, London, 1977.

[63] R.E. Jones, *The Changing Structure of British Foreign Policy*, Longman, London, 1974.

[64] Quoted by John Baylis, '"Greenwoodery" and British Defence Policy', *International Affairs*, Vol. 62, No. 3, Summer 1986, pp. 448, 449.

[65] John E. Woods, *op. cit.* The pseudonym employed by this author would seem to conceal the name of a person who at the time the article was published had sufficiently close official connections with the navy to make the employment of a pseudonym prudent.

[66] James H. Wyllie, *The Influence of British Arms*, George Allen and Unwin, London, 1984, p. 13.

1980s, whose pros and cons are well summed up by Baylis in his 1986 article 'Greenwoodery and British Defence Policy'.[67] Whilst there is nothing to be gained by blowing life into these particular old embers, it is worth remarking that from the perspective of this book, both Greenwood and his opponents are, in a manner of speaking, correct. It is merely that Greenwood's eye seems to have been focused on the operational and development-industrial levels of defence policy, whereas his opponents and the conventional wisdom seem to have had declaratory policy in their sights.

Where British academic commentators on defence do not explicitly show an awareness of the possibility that the declaratory level of policy may not be all that there is to be discussed, they often nonetheless betray an unmistakable uneasiness that this might indeed be the case. Christopher Coker, in his *A Nation in Retreat?*,[68] which on the author's own admission is a polemic on British defence policy written from a strongly pro-NATO standpoint, notices more than once a difference between what British armed forces seem in fact to be capable of and held ready to do, and what (in our terminology) the declaratory policy of 'NATO first' would seem to require of them. Not surprisingly, he is seriously concerned that these discrepancies may be perceived by its NATO allies as evidence of British insincerity of purpose.

In a very similar way, James Wyllie's *The Influence of British Arms*,[69] which is focused on the opposite end of the spectrum from Coker's in being a history - from Suez to the Falklands - of British military activities outside the NATO geographical area, also comes to the conclusion that Britain has paid less attention to its NATO roles than it might have. Wyllie is as keen as Coker to see the imbalance between (in our terminology) the declaratory and the operational put right in favour of NATO.

In many respects the most interesting example of not quite articulated unease comes from Sir James Cable, a former senior member of the British diplomatic service. Writing in 1983,[70] he takes the opposite position to that of Coker and Wyllie, believing Britain has put too many

---

67 *loc. cit.*

68 Christopher Coker, *A Nation in Retreat?*, Brassey's, London, 1986.

69 Wyllie, *op.cit.*

70 James Cable, 'Interdependence: A Drug of Addiction', *International Affairs*, Vol. 59, No. 3, 1983.

eggs in the NATO basket; at the same time, he cannot help doubting whether this is really so, detecting something like our operational level behind what he too calls the 'conventional wisdom' of British defence policy. Cable's 'conventional wisdom' sounds very like the content of our declaratory level of policy.

This attempt to solve the puzzle about whether British defence academics are part of the declaratory circle or disinterested observers seems to have produced the answer that they are both. Where a conscious awareness that the declaratory level of policy may not be the whole picture is absent, there is as often as not evidence that it is unconsciously present. The British academic world is not divided into those who are members of the declaratory circle and those who are not. Rather the case seems to be that many academics are to some degree members of that circle and simultaneously to some degree independent of it. In this they are even closer to senior governmental politicians than we have hitherto suggested.

Inconsistencies and contradictions

Introduction and caveat

Paul Nitze, in his role as the intellectual forerunner of Ball and Snow, and writing for an American and allied audience wanting enlightenment on the niceties of the nuclear declaratory policies of President Eisenhower and John Foster Dulles, writes '[that] the more we can bring our action policy and our declaratory policy into line with each other the more effective both become'.[71] Nitze's ruling seems plain common sense, but it should not be interpreted over-literally. An Australian academic commentator, again with a background in government service, Coral Bell, has pointed out that ambiguity of policy is not always a bad thing and can be productive.[72] In particular she suggests that the traditional external impenetrability of the operational level of United States' nuclear strategy has been in doubt ever since the development of reconnaissance satellites allowed the Soviet Union - during the second

---

[71] Nitze *op. cit.*, p. 198.

[72] Coral Bell, 'From Carter to Reagan', *Foreign Affairs*, Vol. 63, No. 3, 1985, pp. 490, 510.

half of the Cold War at any rate - to receive 'signals' from both declaratory and operational levels. Where these signals did not say the same thing, she argues, this could have been beneficial to United States' policy-makers whenever they may have wished to keep their Soviet opposite numbers guessing. Whilst it is very hard to imagine anything like the same degree of transparency evolving with respect to conventional forces (in the sense that one can tell a great deal more about the probable function of a strategic nuclear launch vehicle from remote photography than would be the case with a destroyer of the Royal Navy), Bell's point retains some force. For according to Desmond Ball, inconsistencies between levels of defence policy can show up in the form of contradictions within levels. Thus it can follow that contradictions within the declaratory level of conventional defence policy may do just the sort of service Coral Bell refers to. However, there is a difference between inconsistencies that have been deliberately fostered or left unattended to simply because they are productive, and those which have emerged and developed as a result of failure on the part of key personnel within the declaratory circle to correct a drifting out of alignment.

The defence review revisited

Bringing the three levels we have identified for British defence policy into line (*pace* Bell) is a job for government politicians of ministerial rank and their advisers, who as a group are uniquely placed in the declaratory circle but with privileged access to the other two circles, and in an important sense members of none. This last attribute is necessary if they are to be effective agents of change and one that they share with some academic commentators on defence.

Interestingly, the process of drawing up the 1990 defence review 'Options for Change' has been the subject of an insider's account. We can learn something from this account in spite of serious doubts as to the genuineness of the review itself; such doubts were not eased by the slightly odd fact that no document as such emerged from the review process, merely a ministerial statement to the House of Commons and a subsequent gloss on this in the ensuing statements on the defence estimates. Something of the flavour and a hint (in the final sentence) of the adversarial nature of what was nonetheless officially described as an 'important defence review' comes from an account of the process involved in producing 'Options for Change', given by a senior Ministry of Defence civil servant, Richard Mottram:

> Essentially the exercise was run at three levels: at the top, close ministerial control by the Secretary of State and his colleagues; a central group comprising all the top management, including single-service Chiefs of Staff, which looked at emerging proposals on their way to ministers; and a small working group of central policy and programme staffs. This approach was designed to ensure close central direction and a defence-wide approach, and to allow for radical options to be addressed on their merits in private, rather than in the newspapers.[73]

This bringing into line of the three levels of defence policy is not practised continuously, although attempts at small adjustments are frequently made - as noted above, no two statements on the defence estimates are exactly alike - and, as we also have already seen, sometimes slightly larger adjustments are attempted without much fanfare. But on the whole, drifting out of line seems to be a slow but cumulative process so that a point is eventually reached when a new defence review cannot be postponed any longer. Barring accidents, then, the new review sets the tone of policy for the next few years.

Other observers have seen defence reviews in a very similar light. Reacting to the Nott review, the then Chairman of the House of Commons Defence Committee, Cranley Onslow, addressing John Nott who was attending as a witness before the Committee in 1981, said:

> ... some of us have felt that defence reviews are an admission of failure. If the object of policy is to keep defence in tune with the nation's needs and resources, if there has to be some kind of great leap in any given direction in the shape of a defence review, that suggests the system has been failing for some time past.[74]

---

[73] Richard Mottram, 'Options for Change: Process and Prospects', *RUSI Journal*, Vol. 136, No. 1, Spring 1991.

[74] 'Ministry of Defence Organisation and Procurement', Minutes of Evidence, House of Commons Defence Committee, Session 1981-82, 11 November 1981, HCP 22-i, HMSO, London.

In fact the trigger for a defence review is often, perhaps invariably, an outside event. A saturated solution of a crystal-forming chemical will not spontaneously start producing crystals unless a foreign object of some kind is introduced to precipitate the process. Defence reviews seem to need a similar sort of 'seed' to precipitate them. A perceived failure of defence policy at the operational level - a use of force gone seriously wrong - is one; the Sandys review in the wake of the Suez crisis would fall into this category. Perhaps this is also true in reverse. A successful use of force can abort any review that may have been in process (see the Falklands and the Nott review) - although statistically speaking this must be a very rare event. A financial crisis is another precipitating factor: the rapidly deteriorating economic situation leading to the devaluation of sterling in 1967 triggered the series of Healey defence reviews. A change of governing party is yet another. The near coincidence of the Nott review with the accession to power of the first Thatcher government fits the bill here - although since a change of governing party is always a result of an adverse electoral verdict on the policies, including the defence policies, of its predecessor, it is not clear that it should be regarded as an altogether distinct precipitating factor.

Bringing the three policy levels back into line in the context of a new or a restated declaratory policy is what defence reviews are for. In terms of public consumption, the consequence of such a review will normally be a statement arguing the mutual consistency (in our terminology) of the three levels of policy now reached or, more probably, now planned for. The invisible, or very largely invisible, prerequisites to such a statement will involve the messy internal Whitehall business of bringing into line the three levels and their circles of actors. Part of this largely invisible process must include the writing and circulation of documentation describing - not of course in so many words - the relative positions reached by the three policy levels during the period since the last review (it has to be admitted that the phrases 'close control', 'small working group' and 'private' in the above quotation from Mottram all too clearly suggest that what he is describing does not quite equate with the role that we have specified for the defence review).

The declaratory level will normally be the only one of the three already existing in a compact stated form or document. The Ball/Snow analysis of United States' nuclear policy derived much of its impact from the fact that they were in possession not only of documentation from the declaratory level, but also had strong evidence for the existence, and some knowledge of the content, of a document from the operational level too, in the shape of the SIOP. Whilst a British version of the SIOP - the plan for the conduct of nuclear war against the Soviet Union - may well have existed during the Cold War (it is a measure of the relative secretiveness of British society that it is impossible to be even 50 per cent sure about this), a British SIOP would be an absurdly simple statement compared to a description of what military commanders understood British *conventional* forces to be for, and how, in the eyes of the development-industrial circle, the requirements of these forces and related objectives were served by the British defence-contracting industry.

Fortunately, operational documentation is not necessary. Instead there is something at least as good - actual experience of operational and development-industrial activity. Since 1970 Britain has fought in two or two and a half important conventional wars, once essentially alone, in the Falklands; once in coalition, against Iraq; and once, with regional help, in the half-war - the 'secret war' - in Oman. Within the development-industrial level, fragments of evidence exist relating to the activities of the British defence-contracting industry in the form of economic data. In addition, supplementary indicators of inconsistencies between the declaratory and the other levels of policy can be obtained from a scrutiny of inconsistencies within each level, but chiefly, since it is the most visible, the declaratory level.

Aims and objectives

In one way this book can be seen as an attempt to reconstruct a representative fraction of what might have been written by government advisers about the operational and development-industrial levels of defence policy in the period since 1981, in preparation for the next review of comparable seriousness of intention and execution to that of John Nott. In other words it is an attempt to provide for Britain's conventional defence policy the equivalent of the SIOP documentation and something else besides. Naturally it will fall significantly short of what official surveys could achieve in terms of access to the facts made available to members of the operational and development-industrial circles. Correspondingly, it will be a deficient reconstruction, with more than its fair share of inferences and educated guesses.

Another way of looking at the book is to see it as presenting a new way to understand British defence policy - a new framework for analysis - in an attempt to cast some light on puzzles about British defence policy which have long troubled the author, and, it may not be too presumptuous to suggest, not the author alone.

The puzzles already referred to in the Introduction have a kind of objective hierarchy. At the top there is the large size of British defence budgets, relative to those of other major European members of NATO, when Britain's apparent contribution to NATO is not correspondingly exceptional. Next there is the puzzle that triggered the author's original curiosity, concerning not only the size of the navy, but also the sorts of ships it possesses. For the years 1970/1990, the period of our study, the navy has been, in terms of the most commonly accepted measure of strength - the total tonnage of warships (including submarines but excluding the four submarines carrying missiles of Britain's strategic nuclear force - the SSBNs) - the second largest in what used to be called the Western world. In 1990 it accounted for 12 per cent of all NATO warship tonnage, making it significantly stronger than the combined tonnages of the next two strongest navies (the French and the German).[75] Its range of large vessels seems particularly difficult to square with the convoy escort duties that it was ostensibly held in readiness for should a comparatively short great-power war occur. Questions about the navy and what it is for soon lead to reciprocal questions about the army and the air force.

At about the same level of importance there are questions outstanding against the British defence-contracting industry. Here again size is the puzzle, or rather size as a function of its apparently sacrosanct status that has protected it from having to tender against international competitors for major Ministry of Defence contracts. At no time in the period between 1970 and 1990 have imports of armaments exceeded 10 per cent of Ministry of Defence procurement, whereas 30 per cent of the demand within the country for manufactured goods as a whole is met by imports. This was doubly a mystery during the 1980s when the prevailing governmental wisdom was to eliminate sacred cows in British industry so that market forces should have freedom to exert their

---

[75] *Report on Allied Contributions to the Common Defense* (A Report to the US Congress by the Secretary of Defense), May 1991, p. A-22.

influence.

These puzzles remain unresolved because they cannot be understood, or cannot readily be understood from within a framework for analysis that has no place for multiple levels of defence policy or multiple circles of actors. Within a framework that embraces the idea of multiple levels, the puzzles immediately fall into a pattern. As we shall see, they can be understood in every case as manifestations of a falling out of alignment of the three levels of defence policy - the declaratory, the operational and the development-industrial. It is to the operational that we now turn.

Section Two: The Army, the Navy and the Air Force

The operational level

The next stage in the enquiry into the nature and extent of the mismatch between the declaratory and the other two levels of British defence policy focuses on the operational level and is composed of two strands, closely intertwined around a core. The core is the defence review initiated by John Nott in 1981 (the chronological mid point in our survey) and the officially published materials relating to its conclusions. One of the strands around the core relies on contemporary accounts of the three wars Britain has been involved in since 1970: in Oman, the Falklands and in the Gulf. The other concentrates on the public utterances of senior serving officers and others positioned on both the declaratory and the operational levels. The ambiguity of such positioning depends upon circumstances. Retired officers, for instance, can speak more candidly on behalf of the operational level than serving officers. However, senior serving officers in particular, whilst their public pronouncements are officially within the declaratory level, will occasionally use 'coded' language which is meant to be read differently within the operational level than from the outside. Even when 'code' is not resorted to, considerable evidence (as we shall see) points to divergent readings of the essence of British declaratory strategy in different branches of the armed services. In fact there are serious inconsistencies, between the navy on the one hand and the army and the RAF on the other. Following Ball's prescription outlined in Section One, inconsistencies at one level of strategy are taken as evidence for inconsistencies between levels, in this case between the declaratory and the operational.

Flexible response

The declaratory role for Britain in NATO was that of contributing to flexible response, wherein NATO armed forces would be ready to resist a Soviet-led conventional attack on Western Europe using only conventional forces, but only for a certain period of time. Beyond this point, NATO forces would be prepared to introduce nuclear weapons into the battle zone and beyond. The exact point in time and in the general progress of the war at which the nuclear threshold would be crossed could not be made clear in advance, since that might constitute an invitation to the other side to apply force with an intensity that kept the war just below this threshold. On the other hand, whilst setting the threshold at a very low level (of NATO casualties, say) would obviate this particular danger, it would court another - the risk of not being believed - since the likeliest consequence of a NATO nuclear strike would be a return of nuclear fire and, almost inevitably, a nuclear war of unbridled intensity which would consume both sides indiscriminately. At the same time, this deliberate absence of clarity was underlined by a policy of nuclear control at the theatre and battlefield level which - particularly where battlefield (that is to say comparatively short-range) nuclear weapons were concerned - seemed to surrender at least some of the responsibility for deciding on a nuclear release to subordinate commanders, even at field level where the chance and confusions of war could be expected on their own to work against restraint.

This lack of clarity cannot be resolved by a simple appeal to the facts. The crude numerical superiority of the conventional forces of the Warsaw Pact, which presumably would have forced all the NATO allies (and not just Britain) to make a quick choice between defeat and nuclear escalation, tells us little. The reason is that the raw figures can be interpreted in different ways. One can plausibly assume that the intrinsic superiority of being on the offence (with the advantages of surprise and holding the initiative), coupled with the advanced technical quality and high inter-operability of their armaments, would make the paper superiority of the East decisive. But one can equally plausibly assume the opposite, citing the inherent superiority of being on the defensive (with its advantages of familiar terrain and short lines of supply), the superior technical quality of NATO armaments and the political disunity and hence military unreliability of Pact forces, all leading to a cancelling out of any numerical weakness on the part of the West.

Army evidence

By the end of the 1970s it had become apparent at the highest political levels in the Ministry of Defence in London that something had gone badly wrong with Britain's NATO strategy. In a nutshell, the navy was ostensibly preparing to fight a long European war, whilst the army and the air force were able to fight only a short one. John Nott's signed introduction to the Statement on the Defence Estimates for 1981 (dated April) refers to 'deep cuts' having been carried out under his predecessors in the procurement of ammunition, fuel and oil and essential spares.[1] Then in his defence review - 'The Way Forward' - of June 1981,[2] Nott refers to the government's intention to procure 'proper war stocks' and to enhance 'war stocks and ammunition to improve the combat endurance of 1(BR) Corps'. In the Statement on the Defence Estimates for 1982,[3] the government pledges that 'we shall devote ... attention to its [the army's] powers of endurance by building up war stocks at every opportunity'. The Statement on the Defence Estimates for 1983[4] seems to imply some progress: (for BAOR) 'war reserves and stock levels are also being increased'. But according to a senior British army officer, General Sir Hugh Beach, writing in 1985 or 1986, four or five years after his retirement from the most senior army equipment procurement post - Master-General of the Ordnance - a war in Europe would last (*pace* the slight hesitation signalled by the question

---

1 *Statement on the Defence Estimates*, Cmnd 8281-I, HMSO, London, April 1981.

2 'The Way Forward', Cmnd 8288, HMSO, London, 1981.

3 *Statement on the Defence Estimates* 1982, Cmnd 8529-I, HMSO, London.

4 *Statement on the Defence Estimates* 1983, Cmnd 8951-I, HMSO, London.

mark) for a 'period of days rather than weeks?'[5] Moreover, it would seem that war stocks were only part of the problem of endurance. According to General Beach:

> Despite the huge outlays on equipment in recent years, the Army can look forward to seven out of its twelve armoured regiments being equipped with Chieftain tanks (1950s technology), perhaps till the end of the century. Six of its artillery regiments still have 105-mm Abbots (of similar vintage), almost useless on the present battlefield (though there are some excellent equipments in the pipeline: the SP70 gun, if it ever comes, and the Multiple Launch Rocket System).[6]

General Beach's wary scepticism was justified. According to an MOD civil servant the SP70 (self-propelled) 155mm howitzer, being built on an internationally collaborative basis between Britain, West Germany and Italy, 'started development in 1973, after much delay, and finally petered out in 1986 with nothing to show'.[7] The Multiple Launch Rocket System (MLRS) was originally meant to go into service with the army in 1984. Delay meant that the equipment did not arrive until 1989, and even by the time of the 1990/91 Gulf war, only 12 units of the MLRS had been delivered to the army.[8]

The war in the Falklands in 1982 neither contradicted nor confirmed what the Nott review had claimed about the endurance capacity of British land and air forces. There certainly were logistics problems. According to General Moore, the Commander Land Forces Falklands Islands, 'By the morning after the final night attacks [before the fall of Port Stanley], some of my batteries were down to only six rounds per gun'.[9] However it does not follow that what stores Britain had available to fight the Warsaw Pact were scarcely enough to fight the

---

5 Hugh Beach, 'British Forces in Germany, 1945-85', in Martin Edmonds (ed.), *The Defence Equation*, Brassey's Defence Publishers, London, 1986, p. 167.

6 *ibid.*

7 Martyn Bittlestone, 'Cooperation or Competition? Defence Procurement Options for the 1990s', *Adelphi Papers*, 250, Brassey's for the IISS, London, 1990, p. 58.

8 *The Economist*, January 26, 1991, p. 27.

9 Major-General Sir Jeremy Moore and Rear-Admiral Sir John Woodward, 'The Falklands Experience', *RUSI Journal*, March 1983, pp. 25, 32.

Argentines. The fact that the Falklands was at the end of a very long line of communications could have meant that shortage of ammunition was the result of an original underestimation of what would have to be drawn from stocks held in Europe, rather than a reflection on the size of these stocks. Even so, Coker, who also refers to shortages of ammunition in the Falklands, goes on to say that the campaign showed up the far more telling fact (from the perspective of the argument advanced here) that the British army was short of trained gun crews.[10] And there is other evidence from the war that reflects directly on important aspects of the stock situation overall.

Thus, referring to US President Reagan's announcement on April 30 1982 of military and economic sanctions against Argentina and his parallel offer of 'materiel' to Britain, Hastings and Jenkins indicate that US aid to the British forces had already been flowing for some time, with supplies including 'Sidewinder and Shrike air-to-air and air-to-surface missiles'.[11]

The missiles referred to would take up too little shipping space in the long supply haul from Europe to the Southern Atlantic for logistics to be a plausible explanation for the shortage the US President was offering to make up. Eddy and his co-authors are a little more specific on the Sidewinder question, stating that the United States agreed at some stage in the campaign to supply Britain with 100 modern units of this air-to-air missile.[12] Freedman adds: 'The Sidewinders were supposed to replenish Britain's NATO stocks but some may have found their way to the South Atlantic.'[13] At any rate the strong implication is that Britain's existing overall stocks of this key air-combat missile were either too small or insufficiently modern or both to allow 100 modern Sidewinders to be painlessly diverted to the Falklands Task Force from stocks.

The Gulf war of 1990/91 confirmed the suspicion that the endurance problem identified by Nott at the start of the 1980s, as having been created in the previous decade by his predecessors, had not been put

---

10 C. Coker, *British Defence Policy in the 1990s: A Guide to the Defence Debate,* Brassey's Defence Publishers, London, 1987, p. 102.

11 Max Hastings and Simon Jenkins, *The Battle for the Falklands*, Michael Joseph, London, 1983, p. 142.

12 Paul Eddy and Magnus Linklater with Peter Gillman, *The Falklands War*, Andre Deutsch, London, 1982, p. 194.

13 Lawrence Freedman, *Britain and the Falklands War*, Basil Blackwell, Oxford, 1988, p. 72.

right by start of the 1990s. A *Sunday Times*[14] report on that conflict claimed that in order for Britain to send one hundred and forty modern tanks from the British army in Germany to Saudi Arabia, as part of its contribution to the coalition war against Iraq, the shortage of spare parts, described by the newspaper as 'chronic', was such as to require the partial dismantling (cannibalising) of all but sixteen of the original fleet, which had totalled 700. In other words the report is saying that in 1990 and for some time before, out of a force on paper of 700 tanks available for war in Europe, only 140 or so were ready for prolonged action even before the first shot was fired. A different report comes to a substantially similar, if not identical, conclusion. According to the *Independent on Sunday*,[15] the British army contributed 160 Challenger tanks to the land war in the Gulf, which were kept operational only by denuding the rest of the army (in Britain and in Germany) of supplies. According to the same source, this left the army 'debilitated', with 'three-quarters of its tanks inoperable because of spares shortages'. Once at war in the Gulf, the combat endurance of the Challenger tank seems to have become a limiting factor on the effectiveness of the British contribution to the land campaign. According to the Statement on the Defence Estimates for 1992:

> The main armament [of the Challenger] was another notable strength but *vehicle reliability* and performance of the fire control system were limiting factors in the tank's overall effectiveness.[16] (emphasis added)

The run-up to the Gulf war was also the occasion for the press airing its worries about the ammunition supply question. Whilst it is always important to scrutinise claims about ammunition with care, these reports were triggered off by a British governmental request to Germany, Belgium and Holland for additional supplies of ammunition for use in the Gulf.[17] The same defence correspondent, Christopher Bellamy, quoted 'defence sources' as saying that there had been a serious ammunition shortage (in the British army) for years and that the army

---

14 *Sunday Times*, 10 November 1991.

15 *The Independent on Sunday*, 10 March, 1991.

16 *Statement on the Defence Estimates 1992*, Cm 1981, HMSO, London, July 1992.

17 *The Independent*, 24 December 1990.

would have been unable to sustain a war in Europe for more than two days. Bellamy added that in order to have enough ammunition to fire on NATO exercises, towards the end of the 1980s British forces had to dip into war stocks.

Confirmation that the Gulf war had thrown light on a real ammunition problem came with a later report that indicated that whereas, given notice, British Royal Ordnance factories could produce ample quantities of 120mm tank shells, they could not produce the extra 155mm shells needed to meet the expected demand on British forces in the Gulf even if given several months' notice.[18] In searching for the latter (presumably because stocks were low) the government was forced to approach its NATO allies. This expedient would hardly have been available in the event of a European war.

Further confirmation comes from General Sir Kenneth Hayr, writing in his capacity as Deputy Chief of the Defence Staff (Commitments) after the conclusion of the Gulf war. As General Hayr points out:

> Government policy is that defence commitments outside the NATO area shall be supported and sustained as far as possible from equipment and stocks procured for the Priority One (*sic*) NATO commitment, so there is no separate 'out of area' stockpile of munitions.[19]

This rules out the theoretical possibility that the ammunition shortages commented on in the press were in some sense artificial and that the shortages referred only to the forces' holdings earmarked for non-NATO purposes. General Hayr admits to shortages of ammunition, but not to the extent of precisely confirming the press reports already referred to (he does not dispute them either). He says:

> There were some particular munitions for which we had to call upon Allies for assistance. With the new Multiple Launch Rocket System we had a brand new weapon in our armoury, and our stockpile was still being built up. For eight inch artillery it was the opposite: this weapon was being withdrawn from service and its stocks were being run down, but it still had valuable utility in

---

18 *The Economist*, January 26, 1991, p. 27.

19 Sir Kenneth Hayr, 'Logistics in the Gulf War', *RUSI Journal*, Autumn, 1991, pp. 14, 18.

> the particular environment of the Gulf.[20]

By way of background explanation for this state of affairs, the General notes that 'the calculations against which stockpiles have been procured for defence in the European Theatre are not necessarily appropriate to a mobile armoured offensive in the desert against dug-in, well-prepared defences'.[21] In other words, the conventional phase of the former would be shorter than the probable duration of the latter. As to what this was estimated to be in the case of the war with Iraq, we are told that Air Chief Marshal Sir Patrick Hine, the joint commander in the Gulf, 'was directed to deploy munitions sufficient for several weeks of battle, plus an in-theatre reserve'.[22] That this 'several weeks' would be longer than the conventional phase of a war in Europe is not perhaps as precise an estimate as one would wish for, certainly not as precise as the estimates anonymously volunteered to the press at the time. For greater precision we turn from army spokesmen to the RAF.

Air force evidence

The conventional phase of the type of war in Europe being prepared for by British and allied forces was less of a ground war and more of the air-land kind, not very unlike, in some aspects at least (but very much less one-sided than) the 1990/91 coalition war against Iraq. Correspondingly, an attempt to probe the anticipated duration of such a war would be incomplete if no attention were paid to air force thinking and planning on the matter. It soon becomes clear that during the past two decades the apparent capacity of the RAF to fight a conventional war against the Soviet Union and its allies of more than a few days' duration was no greater than that of the army.

Wing Commander Parkes, writing in 1976 as an RAF serving officer but on detachment to the USAF[23] and, of course, before the Nott review, questions the validity of the USAF doctrine of deep interdiction (the use of aircraft to attack the military potential of the enemy before

---

20 *loc. cit.*

21 *loc. cit.*

22 Hayr, *op. cit.*

23 Wing Commander Alan Parkes, 'Air Interdiction in a European Future War', *Air University Review*, 1976, XXVII, No. 6, pp. 16, 18.

the latter can be brought to bear on one's own forces) in a European war on the grounds that the conflict would be too short for the enemy's *potential* to matter one way or the other. Accordingly he recommends the husbanding of valuable aircraft for more relevant tasks, including the delivery of nuclear attacks. According to Wing Commander Parkes:

> Although there are many different opinions as to the nature of such a war [as NATO is geared to fight in Europe], there is almost total unanimity regarding its length. A future war in Europe would be a short war. And this must be a prime factor in determining the value of interdiction. For while its effectiveness is undisputed in a long war of attrition, it becomes open to question in, say, a thirty-day war.[24]

Writing in 1979, or a little earlier, about tactical air power in NATO, David Greenwood concludes: 'present attrition estimates [for tactical air] are high enough already, given the expected air defence density in Europe: it is plain to see that compounding survival rates ... means that an air force ceases to exist after a short period'[25]. Writing later, in 1988, the then Deputy Chief of Staff (Operations), Headquarters Allied Air Forces Central Europe (Air Vice Marshal Walker) confirms Greenwood's qualitative claims and strongly indicates that Parkes's thirty-day war was, or had become, a considerable overestimate.[26] After quite reasonably disclaiming any precision for the figures he employs, since they are estimates based on likely Soviet anti-air capacities, and acknowledging that he is in any case constrained to employ unclassified data, the unmistakable impression given by his article is the realism of the calculations he presents or refers to.

The part of Air Vice Marshal Walker's analysis that interests us begins by noting:

> At typical intensive sortie rates a ten per cent attrition could halve

---

[24] Parkes, *op. cit.*

[25] David Greenwood, 'Defence Programme Options to 1980-81', in Mary Kaldor, Dan Smith and Steve Vines (eds), *Democratic Socialism and the Cost of Defence*, Croom Helm, London, 1979, p. 153.

[26] Air Vice Marshal John Walker, 'The Conundrum of Air-Land Warfare', *RUSI Journal*, Summer 1988, pp. 15, 22.

> a force well inside three days, even a five per cent attrition would halve a force within five days. Air commanders - on both sides - will have to remain perpetually conscious of such mathematics. (It is necessary to point out that attrition mathematics do not apply only to aircraft - the units could equally well be ships or tanks.)[27]

We can learn something from each of Air Vice Marshal Walker's two attrition rates, provided we first construct a small model of the air war. A ten per cent attrition rate means that after each group sortie, the force of aircraft is only 90 per cent of what it was at the beginning. It follows that, after seven sorties, the original force is about half of what it was. At the lower attrition rate of five per cent, it takes longer to halve the original size of the force - in fact 14 sorties are required. The only conclusion to be drawn from an application of the model to the above quotation from Air Vice Marshal Walker is that his calculations are based on a tacit assumption of three sorties a day for the aircraft under his command.

The importance of this inference, as we shall see later, is such that confirmation of its reasonableness is needed. This also comes from Air Vice Marshal Walker. Supporting his claim that sortie rates could be the all-important factor, he again uses some 'illustrative figures' to avoid classification problems, and assumes that the enemy has 2000 aircraft whilst NATO has 1000. He goes on:

> But let us say that the enemy ... has adopted a design philosophy which concedes some performance for greater simplicity, reliability and maintainability. Let us say that they can improve on our reliability rates by - say - ten per cent ... Then say that their more simple systems get one more sortie per day out of their aircraft than we get out of ours. Together, they have just changed a two-to-one advantage into a three-to-one advantage as far as practical air operations are concerned.[28]

The preceding extract again only makes sense if NATO aircraft are assumed to have a sortie rate of three a day (in round numbers).

Finally, the Air Vice Marshal turns his attention to the multi-role

---

[27] *loc. cit.*

[28] *loc. cit.*

aircraft - the MRCA Tornado - that constitutes the front line of Britain's air contribution to NATO:

> ... the 30 or so expected sorties from our multi-role machine have to be stocked. To stock fully for both options [air defence and ground attack] with modern weapons could cost two and a half times as much as the aircraft itself. Profligate spending, you may think, when half of that expenditure would, de facto, be unused. Regularly, an apportionment is made; say, one-third air defence to two-thirds ground attack. But if the air defence battle rages longer than planned, by sortie number 11 there is no *multi-role* capability because there are no more air defence weapons. No weapon - no role.[29]

By the same token, by sortie number 21 there are no more ground attack weapons. In other words the conventional phase of the air war is seen as lasting typically for seven days. For as Air Vice Marshal Walker himself points out, there will be no time to wait for replacement aircraft to come through the pipeline. 'From NATO's viewpoint, World War Three is largely a "come as you are" affair.'[30]

The evidence we have so far is that John Nott's worries about the 'combat endurance' of British forces in Germany were far from groundless, and furthermore that the situation if anything proceeded to get worse as time went on, after his defence review foundered on the Falklands campaign.

That the army and the air force, for some time both before the Nott review and after it, seem to have anticipated that a European war and their part in it would be short in duration and that their equipment (especially spare parts, stores and ammunition stocks) pointed unmistakably to the same conclusion does not mean that such a war, had it come, would necessarily have been short. British forces could have found a way of buying time and escaping the dilemma of having either to throw in the towel or to cross the nuclear threshold. The time thus purchased could in turn have been sufficient to allow the British defence-contracting industry to manufacture, if not all, then a fair proportion of the spare parts and other stores needed for a longer war. Whilst the

---

29 *loc. cit.*

30 Walker, *op. cit.*

latter seems somewhat against the balance of probabilities, it is wholly immaterial to the central proposition of this section: that while the army and the air force were preparing for a short European war, the navy was ostensibly preparing for a long one.

The navy

Whilst navies may once have been capable of losing a great power war in an afternoon, their contribution to winning one can only make sense in the context of a war of long duration.

By a remarkable and in some ways fortunate coincidence (for our purposes), the lecture theatre of the Royal United Services Institute in Whitehall, which was the forum in 1988 for Air Vice Marshal Walker's words on the role of British air power in a European war quoted above, became, exactly a fortnight later, the venue for a talk on the role of British sea power in a European war by a senior naval officer, Admiral Sir Julian Oswald, Commander in Chief, Fleet (later to become First Sea Lord).

The two men scarcely seem to have had in mind the same war. Air Vice Marshal Walker's war was seen as lasting days rather than weeks and was recognisably the same war that General Beach referred to and that was envisaged by other commentators and observers quoted above. But Admiral Oswald seemed to be working to a different brief:

> In the short and medium term I believe that it is unrealistic to depend on a redress of the land/air conventional imbalance against NATO Europe; the supreme importance of early, reliable and massive reinforcement will therefore remain ... Let me remind you of the scale of that reinforcement; in the *first* 30 days 300,000 men and 100 Air Squadrons will cross the Atlantic; after *180 days* one and a half million men, eight and a half million tons of freight and three billion tons of POL will have to come across in a total of over 3000 ship sailings and 28,000 trans-Atlantic flights. (emphasis added)
>
> Because enemy submarines could sink reinforcement shipping

> anywhere, we [the navy] have to look to its defence everywhere.[31]

There seems to be very little wrong with Admiral Oswald's projected scale of what the United States could transport in the times given. It took 45 days to ship 200,000 United States military personnel and their equipment to Saudi Arabia at the end of 1990, and whilst this movement was completely unopposed, the distances involved were in many instances in excess of the notional standard 3000 miles of the Atlantic crossing. But what remains puzzling is the assumption that anything like 30 days, let alone six months, would be available.

The Lehman strategy

The last sentence quoted from Admiral Oswald's paper refers to the shift in RN declaratory strategy being considered at the time, away from defending the reinforcement routes defensively - through protecting convoys against naval and air attack in concert with the USN - to defending the same offensively. The new strategy, following the ideas of the then US Naval Secretary Lehman, meant attacking Soviet anti-shipping forces early in the war, before they could deploy into the Atlantic. At first glance it may seem that this has little to do with the estimation of the length of the conventional phase of a European war - but there is a connection.

The difficulty with the Lehman line was the strong possibility that attacking Soviet submarines and Soviet submarine-protection forces in their home waters could not be done selectively. Those submarines destined for anti-shipping roles would be no less vulnerable to this 'forward' naval strategy than those carrying the Soviet submarine-launched ballistic missile forces (SSBNs). Unlike their Western counterparts, which are (or were, during the Cold War) at any one time more likely to be at sea than in port, Soviet SSBNs followed a reverse procedure, with only 20 per cent or less of the fleet at sea at any one time. Thus most Soviet SSBNs would be in the process of deploying from their home bases in the opening period of the war. If the Soviets regarded attacks on the submarines which carried their strategic missiles

---

[31] Admiral Sir Julian Oswald, 'Maritime Concepts of Operation: New Thinking', *RUSI Journal*, Summer, 1988, pp. 10, 14.

as quasi-nuclear, in the sense that such attacks were eroding the Soviet capacity to wage or deter strategic nuclear war - even if only conventional means were employed to carry out the attacks - the Soviet authorities might have decided that the NATO forces had indirectly crossed the nuclear threshold, and responded appropriately. Ironically, then, in following the Lehman line the RN would have assisted in shortening the conventional phase of the war.

There is some evidence that the RN appreciated the possible quandary this raised. Captain M. P. Gretton, writing in 1987 before taking up his subsequent appointment as commanding officer of HMS *Invincible*,[32] and supporting the navy's adoption of the Lehman ideas, nonetheless admits that:

> The British historical experience with offensive ASW [anti-submarine warfare] has been unhappy. In both World Wars convoying had to be reintroduced after severe early losses of shipping, and the Americans needed much persuasion to adopt the same tactic in the later stages of World War II ... Convoying may yet be started after the outbreak of hostilities, but with less assets as more will be involved in forward defence in the Norwegian Sea.

He adds that:

> In the final analysis, there must be a considerable risk in attacking Soviet SSBNs, but one no greater than the risk of using tactical nuclear weapons to counter a deteriorating conventional situation on the Central Front ...[33]

Even so - the awkward SSBN issue apart - it is easy to see why the RN was attracted to the Lehman or Maritime Strategy, for as Captain Gretton points out:

> One of the most important aspects of the Maritime Strategy is the

---

32 Captain M. P. Gretton, 'The American Maritime Strategy: European Perspectives and Implications', in Giles Binney (ed.), *Seaford House Papers 1987*, Royal College of Defence Studies, London, pp. 39, 62.

33 *op. cit.*

declaration that a future United States-Soviet war might be global, conventional and protracted.[34]

Moreover, the Maritime Strategy will have seemed to offer the RN a way of discharging its declaratory role more in keeping with the kind of ships it actually possessed and was reluctant to give up, than the smaller and more lightly armed kind that would be of most use in the escorting of convoys. Admiral Oswald seems to have had fewer misgivings about the future of the convoy than Captain Gretton. The Admiral saw in the Maritime Strategy simply the best chance of winning the 'reinforcement and resupply battle'. Taking the fight to the Soviet navy would force it onto the defensive and reduce the threat it could pose to the all-important Atlantic sea lines of communication.[35]

The Navy view

But none of this obscures the fact that, for Admiral Oswald certainly, the fundamentals of how the RN envisaged its role in a European war - protecting the all-important Atlantic sea lines of communication - had not really changed. In 1984 Admiral Sir William Stavely, the then Commander-in-Chief Fleet, referring to the principal obligation of the RN, wrote:

> The conventional forces on the continent are, we trust, sufficient to hold an attack, but they are not sufficient to fight a full scale continental war: they depend on reinforcement and resupply and 90 per cent of this effort will have to come by sea.[36]

The tiny caveat 'we trust' can be read as signalling that what is apparent to us - a mismatch at the declaratory level between the RN and the other two services - is apparent to others too. This is put beyond all doubt by Admiral Sir Henry Leach speaking in June 1982. Admiral Leach's seniority as First Sea Lord and Chief of the Naval Staff, in conjunction with the recent Nott review and the on-going Falklands

---

34 Gretton, *op. cit.*

35 Oswald, *op. cit.*

36 Admiral Sir William Stavely, 'Maritime Power - Changing Concepts?', *RUSI Journal*, March 1984, pp. 9, 13.

campaign may have emboldened him to speak out and almost abandon 'coded language' for plain speaking:

> Even if all currently envisaged pre-positioning [in Europe] is achieved, 90 per cent of [reinforcement and resupply] will still be seaborne ... Here we could enter the argument of short versus long war. There are those in this country who find it convenient to support the short war scenario but it is a view vigorously refuted by our American allies and not accepted in NATO.[37]

For a more measured naval acknowledgement of the distance between the services on the question of the likely duration of a European war, reminiscent of the observations of Wing Commander Parkes, we can turn again to Captain Gretton:

> The main concern of critics [of the Lehman strategy] is whether NATO could hold out long enough on the ground in the Central Front for a Maritime Strategy to count...[38]

In sum

Admittedly, in reference to the figures themselves, it is just possible to argue that there is a technical overlap between air force and navy estimates of the war's duration. Both Wing Commander Parkes and Admiral Oswald mention thirty days. But the drift of the respective positions of the two services is plainly divergent. For the RAF, thirty days seems a maximum (and the evidence we have on the parlous state of the army's ammunition and spare parts situation would seem to put it firmly in the RAF camp as well); for the navy thirty days seems a minimum. Wing Commander Parkes's implication that it is pointless for NATO to bother to attack the Soviet war potential a few hundred land miles to the east of the front, because there will be too little time for that potential to influence events, leaves NATO's main reinforcements 3000 miles across the Atlantic looking distinctly academic, whatever naval scheme to protect their transit is adopted. And Air Vice Marshal

---

[37] Admiral Sir Henry Leach, 'British Maritime Forces: The Future', *RUSI Journal*, September 1982, pp. 10, 15.

[38] Gretton, *op. cit.*

Walker's homily about the war being largely a 'come as you are' affair scarcely suggests that he envisages much scope for reinforcements either.

The point is not that the air force and the army are right and the navy wrong; things could be the other way about - as we saw above, the European war we are discussing could have turned out to be a long one. Rather the point is that there is an inconsistency within the declaratory level, the sphere in which senior officers operate when making authorised public statements.

Inconsistencies and flexible response

Before proceeding to probe what this inconsistency may mean at the operational and development-industrial levels, it is useful to look at the implications of this inconsistency at the declaratory level of British defence policy. Taking our cue from Coral Bell in Section One, inconsistencies between the declaratory and operational levels of United States' nuclear strategy possibly created a productive ambiguity in the nuclear dealings of United States' political leaders with the Soviet Union. She sees in the development of sophisticated mutual reconnaissance and intelligence-gathering a means whereby at least some of the traditional secrecy of the operational level was lifted. In the present case, it is not even necessary to speculate about the effect on allies and enemies of operational signals. With the Nott review, inconsistencies at the declaratory level of British NATO defence policy became clear for all to see.

In the context of flexible response, a navy preparing for a long war and the other two services preparing for a short one could itself be said to be a productive ambiguity. The army and the air force were playing a tune that sounded right to the ears of Britain's European NATO allies, who on the whole were not keen on planning for a prolonged conventional campaign. The navy's long war planning on the other hand sounded sweet to the United States, whose preferences had long been (ever since, in fact, the Soviets first achieved a reliable nuclear second-strike capability against targets on United States' territory) to de-emphasise the likely role played by nuclear weapons in the defence of its NATO allies. As for enemies, the British ambiguity - seemingly betting about equal amounts on two contradictory scenarios for a war in Europe - helped preserve for NATO that desirable uncertainty in Soviet calculations about when and how and where nuclear weapons might be introduced into a European war.

This would be fine, except for one thing. Coral Bell's productive ambiguity depends on the United States being able to fight a nuclear war either according to its declaratory strategy or in some other way: it had the weapons and planning infrastructure for both. For Britain in NATO, on the other hand, the whole problem is simply that options had disappeared, and with them ambiguity - productive or otherwise. The ability of the army and the RAF 'to hold an attack' for the month plus needed for sufficient reinforcements to arrive from across the Atlantic, escorted by the navy, had seriously diminished or had been allowed seriously to diminish. Their staying power was down to a week or less.

A gloss on Nott

Using our own terminology, a possible reading of the Nott review is as follows. There were few inconsistencies between the declaratory and operational levels of the army and RAF strategies; the gaps that existed were of degree not of kind. But the inconsistency between the declaratory roles of the navy and its sister services had gone beyond the point of being productive and had become pernicious. Outside observers would have gathered that the probability that Britain's army and RAF could fight a war for long enough to make the escorting of reinforcements from the other side of the Atlantic other than academic, was small and declining. In other words the British front-line forces in NATO were operating on the assumption that the nuclear threshold was low, a fact that was becoming generally apparent. The navy on the other hand was in principle preparing for an improbably long European war, whereas operationally - and by definition tacitly - it had conceded that the war would be short and that the operational day-to-day role of the navy lay at least as much outside NATO as within it. Nott aimed to bring an end to the inconsistencies in British declaratory strategy by equipping the army and the air force at least with the option of waging war in Europe at a conventional level for long enough to permit the navy's role of defending the sea lanes to have a bearing on the probable outcome of hostilities:

> We need to set, for the long term, a new force structure which

> will reflect in up-to-date terms the most cost-effective ways of serving the key purposes of our defence effort. The best way of enhancing the deterrent effect of our armed forces, for example in raising the nuclear threshold, is to give more resources to their hitting power and sticking power in combat.

Entirely of a piece with this, the review goes on to state:

> The Government is determined to make more use of reserve forces. We now plan a further substantial expansion of the Territorial Army for use both at home and in Germany ...[39]

Reserves, like other reinforcements, are of more use in a long war than a short one.

Furthermore, if the declaratory and operational levels of naval strategy were in line, the sorts of naval 'assets' required to carry out its NATO operational role of defending the Atlantic route would be different from and cheaper than those the navy possessed and claimed a continuing requirement for. Nott felt that the navy should rely less on surface vessels and more on land-based aircraft and submarine forces for its contribution to allied convoy protection. The cheapness of the new navy would in turn help pay for the munitions, stores, equipment and expansion of the reserves necessary to prolong the sort of war the army and airforce were geared to fight.

However, some of this lay hidden behind a smokescreen. The employment of 'coded' language is not the monopoly of members of the operational circle of actors. The Nott review was itself ambiguous about the fate of residual British interests and military activities outside the NATO area. It spoke of an anxiety to 'exploit the flexibility of our forces beyond the NATO area so far as our resources permit, to meet both specific British responsibilities and the growing importance to the West of supporting our friends ...'.[40] At the same time the review announced severe cuts in the capability of the navy (principally) for engaging in this kind of activity. Only two of the three modern ships (here unblushingly described as 'carriers' - the significance of which we look at more closely below) of the *Invincible* class were to be retained,

---

[39] 'The Way Forward'.

[40] *ibid.*

the *Illustrious* and the *Ark Royal*. The third, the *Invincible*, was to be sold and the older carrier, *Hermes*, was to be phased out as soon as both new carriers were in service. The consequences of this are more profound than they may seem. Because of the need to rotate ships, three carriers permit at least one to be maintained on station in distant waters indefinitely whereas, with only two vessels, this option virtually disappears.[41] Consequently, the case for retaining three carriers is far stronger than the case for retaining two; if the argument for three is lost, the long-term prospects for hanging on to any carriers at all begin to look very dim. In addition, whilst emphasising the value of the Royal Marines ('the Government regards their special experience and versatility as of high value for the tasks both in and beyond the NATO area') the Nott review initially brought forward the phasing out of the specialist amphibious ships *Intrepid* and *Fearless*.[42] Without ships of this sort, it is very difficult to see how the Marines could be of any use whatsoever outside the NATO area. This particular decision, however, was reversed before the outbreak of the Falklands crisis. The review also spoke of reducing the number of the navy's destroyers and frigates, at the time 59, to 'about 50'. The ambiguity of the 'about 50' seems to have been deliberate. According to Grove, the active force of destroyers and frigates would come down to 42.[43] On top of this, plans for the modernisation of older vessels of this type were to be abandoned.

The smokescreen laid down over the out-of-NATO-area aspect of British defence policy included tacit compensation for the planned reduction in naval capability in this sphere by a reference to making as much use as possible of allies and non-military methods of securing British interests 'out of area', as well as some slight expansion of the ability of the army and the air force to contribute.

The Nott review, on this reading, was a classic exercise in seeking to bring the declaratory and operational levels of strategy closer into line. Indeed it also paid some heed to the development-industrial level. In addition to the significant reference to 'non-military methods' of securing British interests, which (as we shall see below) may have had

---

[41] See, for instance, Admiral Woodward, in Moore and Woodward, *op. cit.* p. 25.

[42] 'The Way Forward'.

[43] Eric Grove, 'The Royal Navy: the Fleet Comes Home', in Martin Edmonds (ed.), *The Defence Equation*, Brassey's Defence Publishers, London, 1986, pp. 79, 113.

unintended consequences, the review also announced that the dockyard infrastructure for naval support would be cut by one-third. The considerable inertia represented by the existing quantities and types of warships perhaps made the planned cutting of the naval infrastructure a more telling aspect of the review than might have seemed the case at the time. It signalled, other things being equal, that in the medium if not immediate future, the navy would fall to two-thirds of its pre-Nott strength.

If the war in the Falklands could not by its nature provide crystal clear confirmation that, at the beginning of the 1980s, army and RAF war stocks had been allowed to run down unreasonably far, the success of British arms in the South Atlantic provided sensational evidence of how effectively the navy had clung on to an essentially out-of-NATO-area operational capability in the decade following its supposed abolition. Grove speaks of the war in the Falklands as 'in some ways ... the last demonstration of the capabilities of the old East of Suez Navy before it finally disappeared'.[44]

But even by 1990 it was not clear that the 'old East of Suez navy' had disappeared. The number of major surface combatant vessels was indeed down to 50 from the 1981 figure of 62, with the number of attack or 'tactical' submarines essentially the same at 28, thus constituting a higher proportion overall. But of the major surface combatants, in 1990, 32 per cent were in the higher tonnage class (destroyers plus carriers) better suited for out-of-area tasks than convoy protection duties, compared to a figure of 23 per cent in 1981.[45] The 1980s therefore saw a continuation of the trend towards higher average displacement of Royal Navy vessels which, according to Woods[46], had begun in the 1970s. A 1990 Pentagon assessment of the comparative strength of allied forces shows that in terms of total tonnage of displacement (not including strategic submarines) of major surface and sub-surface combatants, Britain had the second largest navy in NATO, with a combined tonnage greater than the sum of the tonnages of the next two largest fleets (the French and the German). Even when

---

[44] Grove, 'The Royal Navy: the Fleet comes Home', *loc. cit.*, p. 106.

[45] *The Military Balance 1981-1982*, IISS, London, 1981 and *The Military Balance 1990-1991*, IISS, London, 1990.

[46] John E. Woods, 'The Royal Navy since World War II', *Proceedings US Navy Institute*, March 1982, p. 88.

submarines are left out of the reckoning altogether, virtually the same situation applies, but with the size of the British surface fleet now only marginally smaller than the combined size of the next two largest surface fleets.[47]

How can this be accounted for? What, in other words, allowed the gap between the declaratory and operational levels of naval strategy to persist and with it a gap between the declaratory and operational levels of British defence policy as a whole, from the Healey up to the time of the Nott reviews and, if not quite so markedly, beyond? Aside from general observations about the inertia at the operational and development-industrial levels due the long life of ships and the lead times of new construction, there seem to have been four important reasons for the navy's success in having its own way: its global role; weakness at the declaratory level; modest dissimulation; and its links with the development-industrial circle.

The global role

Out of area

The first reason for the navy's success in holding on to an impressive out-of-area operational capability is in some ways the most obvious one of all. Declaratory defence policy in the twenty years since 1970 has never been 100 per cent 'NATO'. The out-of-area part has had two stated components. One has been a Cold War role, with Britain acting (at least arguably) on behalf of NATO objectives outside the NATO area. The other has been the imperial role. Both get a comparatively coherent official description in statements on the defence estimates and elsewhere, although one sort of role can transmute into another, and either is capable of evolving into a third, unstated role - the global one. The global role, which we may take for the moment to stand for the use of British armed forces in the protection of specifically British economic interests outside the NATO area, has never been officially recognised and is used here as a shorthand to encompass operational level activities out of area that do not fall within the official, declaratory out of area

[47] *Report on Allied Contributions to the Common Defense*, A Report to the United States Congress by the Secretary of Defense, May 1991.

pairing of Cold War and imperial roles. The originator of the term is Ian Smart, a former member of the British Foreign Office, who, like us, finds a close association between the global role and the Persian Gulf state of Oman.[48] While the imperial and Cold War roles could be fitted within the 5 per cent or so of the defence budget that was not officially assigned to NATO, the combination of the two together with the global role cannot.

At the start of the 1970s, the chief imperial out-of-area declaratory role for British armed forces was the five-power defence arrangement (FPDA) with Malaysia, Singapore, Australia and New Zealand. A Gurkha battalion in Brunei and a battalion group in Singapore were to be the land component of the British contribution. The combined exercise 'Bersatu Padu' in 1970, involving British naval and air force contributions, was meant to demonstrate the nature of the continuing British connection with the FPDA. In 1971 and on schedule, British bases were officially removed from the Persian Gulf, although frigates remained on patrol. Cyprus accommodated British Vulcan bombers on a Cold War assignment to CENTO, already moribund as an alliance except perhaps for the membership of Iran under the Shah. The navy maintained imperial Beira patrol duties until June 1975 as part of the campaign of sanctions against Rhodesia. But all three services shared in the imperial responsibilities with regard to Belize, while the navy's part in Hong Kong security was not disproportionate.

The new Labour government in 1974 - in pursuit of savings in the defence budget - promised a 'tidying up' of out-of-area activity. But according to Wyllie, amongst others,[49] the brunt of the savings were in fact borne by British forces assigned to NATO duties, particularly those on the so-called Southern Flank. In fact, defence white papers up to the end of the decade show that very little 'tidying up' occurred. True, the British imperial contribution to the FPDA was slimmed down by the withdrawal of the infantry battalion group from Singapore, but the promised withdrawal of the Gurkha battalion from Brunei never materialised. A reduction in the British presence on Cyprus that was also promised by the new government did not occur either, but here

---

[48] Ian Smart, 'British Defence Policy Part I: An International View', *RUSI Journal*, Vol. 122, No. 4, December 1977, pp. 8, 14.

[49] James H. Wyllie, *The Influence of British Arms*, George Allen and Unwin, London, 1984, p. 89.

there was an obvious reason why. The coup in mid-1974 by the Greek Cypriot National Guard and the subsequent Turkish invasion of the island meant that instead of being reduced, the two British Sovereign Base Areas on the island were reinforced, mainly by the RAF. The official statement in 1974 that no British forces were to be committed to a CENTO Cold War role did not prevent two navy frigates and an RFA supply ship from taking part in a naval exercise in the Gulf and Indian Ocean in 1978 - MIDLINK 78 - under CENTO auspices, with the involvement of Iran. The decision in 1974 to halt the specific imperial assignment of frigates to Caribbean duties does not seem to have been carried through either, except perhaps in the letter. Instead of two frigates, there seems to have been only one, but its presence was apparently supplemented by other vessels, including submarines, from time to time.

By late 1979 a 'snapshot' of the navy taken by Coker showed that one-quarter of the navy's 'front line' ships and one-half of the support fleet to be outside NATO waters.[50] This seemed to Coker, quite reasonably, to signal a considerably larger involvement of the Royal Navy in out-of-area business than was apparently compatible with (in our terminology) declaratory policy. To explain this situation, we need to look for reasons other than declaratory imperial and out-of-area Cold War tasks.

Global role hinted at

Moderately helpful clues as to the nature of the global role are scattered here and there within statements on the defence estimates. According to the 1971 Statement on the Defence Estimates, the government recognised 'that British interests and responsibilities are not confined to the NATO area' and claimed that 'there is a recognition that British political and trading interests are world-wide and they can only flourish in stable conditions'.[51] Out-of-area 'stability' was to prove an enduring theme for almost twenty years.

In 1973 the Statement on the Defence Estimates says ' ... Britain continues to play her part, within her resources, in countering threats to stability outside Europe. For without that stability Britain's world-wide

---

50 Christopher Coker, *A Nation in Retreat?*, Brassey's, London, 1986, p. 75.

51 *Statement on the Defence Estimates 1971*, Cmnd 4592, HMSO, London.

political and trading interests cannot flourish.'[52]

The coming to office of Labour governments between 1974 and 1979 meant that even brief declaratory accounts of defence policy out of area were dispensed with.

But the new Conservative government in 1979 reinstated the tradition. The Statement on the Defence Estimates for 1980 insists that British defence resources should be used to 'help protect our own and more general Western interests over a wider area including outside NATO'. It goes on to speak reasonably enough of the importance of defending dependent territories and claims, less reasonably, that with Britain depending on the developing world for many raw materials, securing trade routes is of vital importance, as is its interest in the stability of countries with whom it trades. It adds that the Soviet Union has an increasing capability to intervene in the Third World.[53] What is not acknowledged is that very many Western countries depend on trade routes and raw materials, including (unlike Britain) oil, from the Third World, and whilst a number enjoy higher levels of national income than Britain, few appear to find it necessary to spend resources on military activity to safeguard these supposedly vulnerable trading interests.

The Statement on the Defence Estimates for 1983 essentially repeats what had been said the previous year about out of area, although it makes the interesting additional observation, first signalled in the Nott defence review, that Falklands-type intervention was not the only way to secure British interests, suggesting as an alternative in the supply of arms and the provision of military training.[54]

The 1984 Statement lapses into uncharacteristic ambiguity concerning out of area, about which it is in any case very brief. In stating that the threat 'we face' is not limited to NATO territory, the 'we' could refer to Britain or to NATO.[55] By 1985, however, all the old certainty and phraseology returned. Reference is made to Britain's strong out-of-area interest in the stability of trading partners and in securing supplies of oil vital to Britain's economy and that of other Western states. The importance of keeping open key trade routes is also stressed, together with a claim that it is in the general Western interest that Britain has a

---

52 *Statement on the Defence Estimates 1973*, Cmnd 5231, HMSO, London.

53 *Statement on the Defence Estimates 1980*, Cmnd 7826-I, HMSO, London.

54 *Statement on the Defence Estimates 1983*, Cmnd 8951-I, HMSO, London.

55 *Statement on the Defence Estimates 1984*, Cmnd 9227-I, HMSO, London

military presence in Cyprus and the Gulf. The old Nott contention about actual military intervention not always being necessary when arms supplies and military training programmes could sometimes achieve just as much is repeated.[56] It is not pointed out anywhere that by far the most important British trade routes were actually across the English Channel and the North Atlantic (by 1985 virtually 50 per cent of Britain's trade by value, imports and exports combined, was with the European Community alone), nor that Britain's supplies of oil have to travel no further than half the width of the North Sea, not to mention that interruption in Middle East supplies of oil would not directly hurt Britain, since it had been a net oil exporter since 1980.

The 1987 Statement evokes an interesting change of key.[57] We have already quoted it in Section One as an example of how declaratory policy can change rapidly. We are told that 'none of our [British] out of area garrisons is constituted with Soviet aggression in mind or even aggression by proxies of the Soviet Union'. In other words the out-of-area Cold War role is defunct. The lameness of the reason given why nonetheless there continue to be such garrisons - a British interest in 'regional stability because of our dependence on overseas commerce' even though '80 per cent of our visible trade is with other developed countries'- may be disingenuous. 'Commerce' hints at more than trade, and the qualifying adjective 'visible' has the effect of directing the mind to the possibility that what is true of visible trade may not be true of the invisible variety.

The 1988 Statement permutates some of the traditional reasons given over the years for British out-of-area activity: the old imperial role of protecting dependent territories; the promotion of stability in areas where Britain has important economic interests; and a new task of protecting British merchant shipping in the Gulf.[58] The reference to 'economic interests' rings far more convincingly than earlier references to trade, but why Britain has to defend its share of merchant shipping in the Gulf when neither Japan nor Germany found it necessary to do so is not explained.

The 1989 Statement is particularly brief on out of area, saying only that Britain continues to promote stability in areas important to 'us and

---

[56] *Statement on the Defence Estimates 1985*, Cmnd 9430-I, HMSO, London.

[57] *Statement on the Defence Estimates 1987*, Cm 101-I, HMSO, London.

[58] *Statement on the Defence Estimates 1988*, Cm 344-I, HMSO, London.

the West as a whole'.[59] Why Britain - comprising an even smaller proportion of the Western economic system in 1989 than it had in 1980 when the annual Statement made the same point - should unselfishly shoulder this task is not explored.

Global role defined

The global role can now be defined with more precision, a role that would have made complete sense to Captain Mahan who believed that through their contribution to trade, navies could be made to pay even in peacetime. In declaratory terms, Britain withdrew from the Arab shore of the Persian Gulf in 1971. Operationally, British forces remained. The ambiguity generated by this contradiction was convenient, for very shortly after the formal British departure, it came to suit the traditional, semi-feudal Arab governments along the Gulf to pretend that their security did not depend on the intimate protection of outside forces. The different lessons of Iraq and Jordan - that it pays even traditional governments to placate nationalist movements both at home and in neighbouring territories - were appreciated quite early by Kuwait and were gradually borne in even upon the smaller sheikhdoms of the Arabian peninsula, whose first instinct on being informed of the British intention to withdraw had been to ask for a change of mind. It also paid Britain to remain involved in the security of the Arab Gulf states, because whilst the visible part of British trade - with the exception of the arms trade - was increasingly turning towards Europe (all the more so with British North Sea oil due to come on stream), the invisible part was not. The dramatic rises in the price of oil that occurred during the 1970s not only made North Sea production economically sensible, but it also meant the transfer of resources from oil consumers to oil producers on a scale too great - in the sense that short or medium term domestic absorption of the inflow of funds was out of the question - to allow the underpopulated oil-producing states in the Gulf to do much more than place the sums in question in interest-bearing deposits and equities in the Western world.

During 1974-75, Kuwait and the United Arab Emirates had a combined (roughly equal) oil revenue of about $13 billion, up hugely

---

[59] *Statement on the Defence Estimates 1989*, Cm 675-I, HMSO, London.

from the $2 or $3 billion of the previous year.[60] How much of this money and that of immediately preceding and succeeding years found its way to the City of London is not clear. As Peter Mangold, a former member of the Research Department of the British Foreign Office writing in 1975 about Anglo-Kuwait relations in the 1950s and 1960s, all too accurately points out, 'banking secrets ... can be more closely guarded than military ones'.[61] Nonetheless, in a shrewd analysis which is applicable not only to the events of thirty and forty years ago, he maintains that in 1963 the size of Kuwait holdings in Britain amounted to one-third of total British reserves at that time. It is true that Kuwait took steps from 1963 onwards to diversify its holdings, and subsequent devaluations of sterling may have given further encouragement to such a policy. Even so, with the accelerating rise in the oil income of Kuwait and the Emirates beginning with the oil price increases of the final quarter of 1970, boosted further by the 'participation' agreements arrived at between the Gulf Arab oil-producing states and the oil companies in 1972 and then surging with the Middle East war of October 1973, Britain was obviously not overlooked as an investment home.

If, as seems likely, the surge in the oil incomes of Kuwait, Qatar and the United Arab Emirates (UAE) did produce a similar surge in the funds deposited by these states in the City of London, this was simultaneously a boon and an embarrassment to Britain. The boon was that the value of sterling was underpinned by these large foreign holdings; the embarrassment was that the purely economic attractiveness of placing funds in Britain was not in itself likely to be so great as to prevent new radical governments in these countries (coming to power as a result of internal or external forces) from shifting the funds elsewhere. Accordingly, a continuing British military involvement aimed at sustaining these friendly governments in power would pay, provided that disproportionate costs were not incurred. Writ small, the same philosophy seems to have applied to the British military presence - essentially a brigade of Gurkhas - in the Far East oil-producing but underpopulated sultanate of Brunei. After the first steep rise in oil

---

60 *Strategic Survey 1973*, IISS, London, p. 35 and *Strategic Survey 1974*, IISS, London, p. 29.

61 Peter Mangold, 'Britain and the Defence of Kuwait 1956-71', *RUSI Journal*, Vol. 120, No. 3, London, September 1975.

prices of the early 1970s, no more was heard of the British intention to withdraw these forces. What had begun as an imperial role in Brunei thus changed to a global role. In 1989 the size of Brunei's financial reserves (chiefly in the form of financial assets held in Britain and elsewhere) was estimated at $25 billion.[62] In the same year the financial assets of the UAE were estimated at $40 billion,[63] whilst Kuwait's overseas holdings of financial assets were put at $85 billion, including 'large holdings' in Britain (although with the British share of the total probably smaller than that of the United States or Germany).[64] British official reserves of foreign currency in 1988 were $33 billion.[65]

Oman

The pivot of continuing British involvement in the Gulf after the official withdrawal of British forces was Oman. Oman is not itself a particularly important producer of oil - it did not begin exporting oil until 1967 - but it is strategically placed. It borders the United Arab Emirates and Saudi Arabia, and, to the south, the Yemen. More significantly its Masandam peninsula (itself an Omani enclave within the United Arab Emirates) commands the mouth of the Persian Gulf. It is sparsely populated - educated guesses put the total native population at about one-half to one million[66] - and the country is divided in two by 500 miles of 'empty quarter', to the south of which is the Dhofar region with a population somewhere between 150,000 and 300,000. The Dhofar is about the size of Wales, and since 1965 had been the focus of an uprising by the People's Liberation Army controlled by the Popular Front for the Liberation of Oman. The war, which originally seems to have been a separatist-inspired uprising against the uneven development of the two parts of the country, took on more revolutionary overtones with the British withdrawal from Aden in 1967. Aden and its hinterland became the South Yemen - the People's Democratic Republic of Yemen

---

[62] *Financial Times*, May 11, 1989.

[63] *Financial Times*, April 14, 1989.

[64] *The Times*, February 24, 1989.

[65] *OECD Survey United Kingdom 1990/91*, Organisation for Economic Cooperation and Development, Paris, 1991.

[66] Penelope Tremayne, 'End of a Ten Years' War', *RUSI Journal*, Vol. 122, No. 1, March 1977, p. 44.

- whose government began to give the Dhofari rebellion support. An imperial role, with increasing Cold War aspects - the PDRY efforts in Oman were supported 'alternatively by Russia and China later by Cuba'[67] - was developing into a global role.

The decisive British counter-stroke was the removal of the ruler of Oman, Sultan Said bin Taimur, in 1970, and his replacement by his British-educated son, Sultan Qaboos, whose thinking seems to have been more to Britain's liking. Sultan Said had also owed his position to Britain, which had intervened militarily in his support in 1957 to put down an uprising led by the Imam of Oman.[68] The quelling of the uprising, in which the Imam had received backing from Saudi Arabia, was reflected in a change of name for the country. What had been 'Muscat and Oman', signalling a country with two rulers, the Sultan in Muscat and the Imam in Oman, became simply Oman. But Sultan Said had proved a disappointment to Britain and was now dispensable. In its smaller and more gentlemanly and humane way - Sultan Said was removed to Britain where he died (in 1972) in not uncomfortable exile - this was a British version of the American ejection of Vietnam's Diem in 1963. The other difference was that it was successful.

Under British direction, the new Sultan began a programme of economic development in parallel with a military offensive against the rebels. The military operation was commanded by British military officers from all three services. The overall commander in the final phase of the war was the British General Kenneth Perkins, upon whose own account of the war this brief description is based.[69] There were detachments of army, Marine and RAF personnel in Oman, but the bulk of the fighting was done by British-officered Omani forces, ultimately about 10,000 strong, supplemented by about 3000 troops from Iran and a further 1000 from Jordan. Additional fighting was carried out by the so-called *firqa*, an irregular force made up of defectors and others, under the direction of the British Special Air Service. The campaign did not go entirely smoothly and on two occasions it looked as though the

67 Major General K. Perkins, 'Oman 1975: The Year of Decision', *RUSI Journal*, Vol. 124, No. 1, March 1979. At the time of publication General Perkins was Assistant Chief of the Defence Staff (Operations) in the Ministry of Defence, London. In 1982 he joined British Aerospace.

68 Wyllie, *op. cit.* , p. 59.

69 Perkins, *op. cit.*

war might spread. In 1972 the PDRY was invaded briefly by its northern neighbour the Yemen Arab Republic with assistance from Saudi Arabia (now on the side of the British), and in 1975 the British-led forces in Oman carried out artillery and air strikes over the border in the PDRY. Nonetheless, by the end of 1975 the war was essentially over, although border incidents with the PDRY continued until 1987. The final settlement came in October 1988, with the official visit to Oman of the PDRY President Al Attas.

Britain's involvement in the 'secret war' in Oman was supplemented by a continuing 'presence' in other parts of the Gulf in all but name. According to successive statements on the defence estimates, between 1971 and 1976 the army carried out exercises in the UAE, the RAF made regular exercise deployments to the area and the navy regular visits. With the end of the war, British land forces in the area seem to have dwindled to groups of army personnel engaged in training Omani and other forces of the smaller states along the Arab littoral of the Gulf, but naval activity remained at a high level. In 1976 the navy deployed a group of nine ships to the Gulf and Indian Ocean and in 1978, as we have seen, it deployed two frigates and RFA support to the MIDLINK CENTO exercise in the same area, while also sending survey ships to the Gulf.

The lull in Gulf politics, inaugurated in 1975 not only by the defeat of the Dhofar rebellion but also by the 'reconciliation' agreements in March and June of that year between Iraq and Iran, which suspended both their quarrels over territory and Iranian help to Kurdish separatists, ended in February 1979 with the overthrow of the Shah. The Iraqi invasion of Iran in 1980 and the new instabilities in Gulf politics led to a further surge in the price of oil. 1980 was the year when British North Sea oil production first exceeded British oil consumption, but this did nothing to extricate Britain from its status as a home for the surplus funds of the underpopulated oil producers of the Gulf and elsewhere; in fact higher oil prices meant the reverse. Combined oil revenues for Kuwait and the UAE in 1985 were in excess of $20 billion (see Table 2.1, p. 87). The British response was to reinforce its discreet military involvement in the Gulf, and this gathered pace during the 1980s.

Successive statements on the defence estimates provide the details. From October 1980, two British warships were put on station in the Gulf as the so-called Armilla patrol. This rose to three ships and four minesweepers by the end of 1987. The same year also saw a high plateau reached by other sorts of British military activity in and around the area. Six hundred British servicemen from all three services were by that time on loan to all governments on the Gulf littoral (excepting Iraq and Iran). Company level training exercises took place in Jordan. Training deployments of Tornado aircraft were made to the UAE. Nimrod maritime aircraft carried out training missions from Oman. At the end of 1986, in November, the exercise Saif Sareea ('Swift Sword') took place in Oman with 4750 British service personnel combining with the forces of the Sultan. At the end of 1987, a different but highly suggestive sort of exercise took place in Scotland. Exercise Purple Warrior was a joint service exercise designed to test the ability of British forces to capture a fictitious island group 1500nm from the British mainland. This was the largest amphibious exercise carried out by Britain since 1945. Twenty thousand servicemen were involved, together with thirty-seven ships.

Cyprus and Brunei

This discreet build-up of British activity, designed to bring reassurance to the rulers of the small Gulf states, was accompanied by a parallel intensification of activity on Cyprus. This had little to do with the political situation on Cyprus as such (and is not to be confused with the British contribution in the form of an army and an army air corps contingent to UN forces - UNFICYP - on the island). The continuing Greek-Turkish stalemate over the future of the island probably makes it easier rather than harder for Britain to retain its sovereign (air) bases there and is consequently not a state of affairs that Britain would be unreservedly happy to see altered. Whilst it is difficult to make out whether British forces, chiefly air, on Cyprus are declaratorily assigned to NATO duties or something else (declaratory policy not being consistent on this point, Cyprus intriguingly and uniquely oscillating three ways between being 'imperial', 'Cold War' and, in our terminology, 'global'), Cyprus is an excellent operational base for British air activity in the Gulf. The high plateau of British military activity in the Gulf in 1987 was mirror-imaged in Cyprus. The reference in the Statement on the Defence Estimates[70] to military activity on Cyprus in 1987 is more extensive than usual. It speaks of

---

[70] *Statement on the Defence Estimates 1988*, Cm 344-I, HMSO, London.

Tornado and Phantom detachments as well as Shackelton, Buccaneer, Canberra, Phantom and Jaguar aircraft training detachments on the island. During the coalition war against Iraq in 1990/91, Cyprus was home to 4000 British servicemen and a base from which 10,000 air sorties were flown. At the war's end, Air Vice Marshal Sandy Hunter was quoted as saying: 'As long as there is a British presence in the Gulf, we [the British forces on Cyprus] will have a part to play'.[71]

The 1980s did not see Brunei neglected either, although the official reports chart a more haphazard intensification of British military activity in the comparatively peaceful Far East than was the case with respect to the Gulf. The Five Power Defence Arrangements became the focus for tri-service exercises in the region. 1984 saw three British warships exercising in the South China Sea; 1988 witnessed a combined maritime and air exercise with the 'largest ever RAF contribution'.[72]

The navy pays its way

In assessing the proportionate or rather disproportionate part played by the navy in the British out-of-area role, the Nott review again provides the central clue. In order for the navy to perform these tasks a distortion was induced. A navy designed to carry out its declaratory NATO tasks should arguably have had more small escort vessels and submarines and relied far more upon aircraft stationed on the British mainland than it currently did. Such a 'Nott' navy would be much less useful in out-of-area tasks. No comparable distortion was involved in army and RAF contributions out of area. Armies and air forces have to exercise somewhere; moreover, the actual cost of peacetime British land and (possibly some) air contributions in the Gulf and Brunei seems to have been largely paid for on a cash basis, in the 1980s at least, by the host governments. The 1985 Statement on the Defence Estimates speaks of 80 per cent of the cost of providing loan service personnel (400 servicemen is the number given) to all the Gulf littoral states (excepting Iran and Iraq) being met by the countries 'helped'.[73] The navy and the arguable distortion in the navy, in terms of the numbers and kinds of ships it retained at its disposal, received no specific compensation.

---

71 *The Times*, 2 October, 1991.

72 *Statement on the Defence Estimates 1989*, Cm 675-I, HMSO, London.

73 *Statement on the Defence Estimates 1985*, Cmnd 9430-I, HMSO, London.

Rather, in this global role, it contributed more genrally to Britain's invisible trading position. So one reason for the navy's success in going its own way was simply that it paid.

Weakness at the declaratory level

Friend at court

The second reason for the navy's success was a decade of weakness within the declaratory circle of actors at its topmost - Cabinet - level. The new Heath government of 1970 at first held out some hope (to the navy) of officially reversing the 1968 decisions of its Labour predecessor to withdraw from East of Suez by 1971, but the real explanation of the navy's success was weakness at the declaratory level, as a consequence of the preoccupation of the Heath government and its Labour (Wilson and Callaghan) successors with persistent domestic economic problems. Somewhat ironically, the effect of being distracted in this way was an actual rise in real spending on the navy during the Wilson and Callaghan governments. According to Grove, in 1978/79 as much was being spent on Navy General Purpose Forces in real terms as had been spent in 1966/67.[74] Weakness at the declaratory level of defence policy was reflected in the turnover of political personnel at the top. In the 1970s Britain had six defence ministers (three Conservative and three Labour) compared to only three in the previous decade. To declaratory weakness may be added an element of luck, in that the coming to prime ministerial office in 1976 of James Callaghan seems to have given the navy a 'friend at court'. Grove describes Callaghan as 'traditionally pro-Navy',[75] while Callaghan's autobiography provides strong circumstantial evidence of a sentimental attachment to the navy and its traditions.[76] Callaghan himself had served in the navy during the Second World War as an Ordinary Seaman; he was proud of that

---

[74] *ibid.*

[75] Grove, 'The Royal Navy: the Fleet comes Home', *loc. cit.*, p. 104. It is not absolutely clear what Grove means. He could be saying that Callaghan as a politician always gave the navy the benefit of the doubt as a consequence, probably, of his wartime naval service. Alternatively he may be referring to a traditional belief presumably in naval quarters that Callaghan had a soft spot for them.

[76] James Callaghan, *Time and Chance*, Collins/Fontana, London, 1988.

and of the fact that he had been identified as officer material shortly before the war's end. Not only that, his father and maternal grandfather had also both served in the navy below deck. In addition, in the first Attlee administration Callaghan held a junior ministerial appointment at the Admiralty; he also owned to subsequent disappointment at not getting the First Lord's job when it fell vacant. In his autobiography the navy gets three entries in an admittedly minimal index: the other two service branches get none.

Time and chance

The 1980s produced slightly more stability at the declaratory level, since there was no change of governing party. Even so, there were five Ministers of Defence between 1980 and 1989, also the turbulent nuclear politics of the earlier part of the period, with the decision to replace Polaris and the more domestically divisive plan to station US cruise missiles in England, had a distracting effect. Indeed the Nott review itself forfeited some intellectual coherence as a result. There was something not quite right about a review intent on seeing the nuclear threshold in a European conflict raised and reliance on nuclear means of defence correspondingly reduced, with its logical emphasis on reserve forces and proper stocks of conventional spares and ammunition, presented at a time when the government was going out of its way to emphasise the paramount importance of the nuclear element in British defence policy, from the strategic (Trident) to the theatre (cruise) to the battlefield (the Lance replacement). And luck was again on the side of the navy. The Falklands War created (from the navy's point of view) helpful confusion as to the appropriateness of the Nott review, while the victory for British arms gave more power to the navy's elbow. Precisely how much the Falklands War contributed to taking the edge off the effects of the review is difficult to make out for, as we have seen, Nott had already begun to retreat in the matter of specialist amphibious vessels shortly before the war broke out. The losses sustained in the Falklands (two Type 42: *Coventry* and *Sheffield*, and two Type 21: *Antelope* and *Ardent*, were sunk) and the coming into service of replacement vessels may not have increased the size of the navy much over what it would otherwise have been, but they helped reduce the average age of its ships.

Inexact terminology

The third of the four reasons for the navy's success in going its own way was that it was careful to use 'coded' language so as not to flaunt the fact that it was intent on retaining an East of Suez or out-of-area operational capability long after the Healey reviews had apparently abolished it - symbolised, as we have seen, first in the doing away with aircraft carriers and secondly with what the air force had intended to put in their place, F-111 aircraft. Accordingly, the navy simply began to mis-describe their ships to the outside world. Thus the 1970s' fleet sounded reasonably like the sort that would be required to protect the Atlantic sea lines of communication during a 'Third Battle of the Atlantic', whereas in reality it contained ships, old and new, that were well suited, perhaps even better suited, to out-of-area roles. Poignant evidence comes from a young British naval officer Lieutenant David Tinker, who died in the Falklands war, writing in April 1982 to his wife from Ascension Island, en route with his ship *Glamorgan* to the South Atlantic:

> I notice that the Navy is upgrading everything. *Invincible* is called a carrier, not a 'through-deck' cruiser. We are called 'light cruisers' not 'destroyers', etc. This is in fact the correct terminology, and I'm glad they are using the proper terms at last.[77]

In fact the forbidden word 'carrier' to describe the *Invincible* class had begun to appear on official documents in 1979, although the vessels were still more often referred to as 'through deck' or ASW cruisers. But by the 1980 Statement on the Defence Estimates, the word 'carrier' was everywhere. Coyness as to the appropriate description for surface vessels like the light cruiser *Glamorgan* had become and remained more deeply ingrained. Seemingly, the pretext that *Glamorgan* and the rest of the County Class (with a displacement of 6200 tons) together with Type 42 and Type 22 vessels (with displacements of 4100 and 4000 tons respectively) could be thought of more or less interchangeably with the Leander class vessels (displacing 2800 tons) under the heading

---

[77] Hugh Tinker (compiler), *A Message from the Falklands*, Junction Books, London, 1982, p. 162.

'destroyers and frigates' proved too useful a camouflage in domestic battles to be readily discarded. In fact, the word 'cruiser' did not come out of purdah until 1992 when the official publication *Defence Statistics* expanded the standard 'destroyers and frigates' heading in its compilation of British naval strength to include this class of vessel.[78]

It was not simply that the navy was using coded language in order to disguise its success at retaining older vessels and with them (as is suggested here) older roles. There was indeed something of that in the redesignation of the *Hermes* and *Bulwark* 'East of Suez' commando carriers as 'through deck' ASW cruisers; but the navy was acquiring new equipment as well about which it also wished to remain discreet. The *Invincible* carrier - at that time still officially described as a 'through deck' ASW cruiser - was ordered in 1973, and two more of the same class, together with appropriate short take-off and landing Harrier aircraft modified for naval purposes (Sea Harriers), were ordered shortly afterwards. According to Woods, however, this was not all (note how the slipperiness of the terminology can make the account of even a close and well-informed observer of the RN less than immediately comprehensible):

> Ironically, the defence reviews of the mid-1970s actually improved the Royal Navy's combatant carrier capability ... the *Hermes* and the *Bulwark* would be reequipped and redesignated ASW carriers from the commando role. These interim through deck cruisers first increased then maintained active carrier strength at two units, despite the over-publicised departure of the *Ark Royal* at the end of 1978. More significantly, perhaps, construction of a new generation of Type 42 air defence, guided missile destroyers and two new Type 21 and Type 22 classes of large, helicopter-equipped frigates had begun.

Woods concludes:

> After a drop to 78 major active units in 1972-73, total [Royal Navy] strength fluctuated between 79 and 85 for the rest of the decade [1970s]. Although superficially a small ship navy, the new fleet's vessels had a rather higher average displacement than

---

[78] *Defence Statistics*, 1992 edition, Government Statistical Service, HMSO, London.

> previously, because of the new escorts and submarines in the 4000-ton category. The Royal Navy that entered the 1980s was for its size probably the most modern and sophisticated in the world.[79]

It is not intended to exaggerate the success of the navy in retaining an operational capability far beyond its declaratory role. For one thing, there is a great difference between *Invincible*-class carriers and the mammoth nuclear-powered aircraft carriers of the US navy. Perhaps the former should have the honest designation of light carrier, by analogy with light cruiser. Moreover, the Falklands campaign showed up serious deficiencies in the navy's ability, as an out-of-area fleet, to obtain early warning of air attack since light carriers seem not to provide an adequate platform for the sort of aircraft necessary to perform this role. But it is hard to imagine much dissatisfaction within a navy that had been told at the end of the 1960s that its East of Suez role would disappear by 1971, and then being able in 1982 to put two light aircraft carriers, three light cruisers, five Type 42 and three Type 22 vessels all of the 4000-ton category, plus eighteen other frigates and destroyers into a successful effort to recapture the Falkland Islands.

The Falklands war, won against a far from negligible opponent at over 7000 miles distance from home base, confirmed the accuracy of Woods's judgement that the navy 'was for its size the most modern and sophisticated in the world'.

A domestic ally

The final reason for the navy's success in resisting attempts to confine it within the requirements of a 'NATO first' strategy is closely connected with the first reason and the global role. The navy had an ally. This was the development-industrial level and its circle of actors, made up chiefly of the British defence-contracting industry.

The British defence-contracting industry has customers for its products overseas as well as at home. These customers are typically in regions of instability, since naturally this is where demand for armaments is greatest. The navy's operational out-of-area role in the Arabian

---

[79] John E. Woods, 'The Royal Navy since World War II', p. 88.

peninsula thus not only paid in helping to ensure that oil money was not precipitately withdrawn from London, but also in ensuring that regional orders for armaments, themselves funded by oil money, were in fair measure also placed in Britain. In the case of Oman (or Brunei), it is almost inconceivable that orders for armaments should have gone anywhere else. But not every state in the peninsula is quite in the category of Oman, described perhaps a little incautiously as late as 1990 as 'generally regarded as still something of a British protectorate'.[80] Even so, since the bulk of all international transactions involving the flow of armaments are between allies (NATO being a classic example of this: pledges of military support, formal or informal, stimulate the supply of armaments rather than substitute for it), Britain's informal alliance network in the Gulf should by and large mean orders for arms flowing almost automatically. Moreover, the discreet British presence in the Gulf in the 1970s and 1980s gave the defence-contracting industry some insurance against regional instability becoming so great as to jeopardise the completion of contracts for the export of arms. The process of filling such contracts is by no means quick when advanced (and expensive) weaponry is being transferred to countries with inadequate domestic physical and human infrastructures. Remedying the deficiencies in infrastructure may itself be highly profitable, but it is invariably a slow business and major sales of arms cannot proceed without it. The fall of the Shah of Iran left British arms exporters all too aware of the danger of 'counting chickens'. Too little instability and there are no orders for arms at all; too much and the completion of contracts becomes problematical. As things turned out, Saudi Arabia, Britain's ally in the secret war in Oman, quickly replaced the Shah, another ally in the same war, as Britain's chief customer for arms.

The alliance had another, less tangible and less important side. To the extent that it may be said that the defence-contracting industry in Britain is preparing for war, it is for the same sort of war that the navy, in declaratory terms, was preparing for - a long great-power war. A long war of this kind means that it makes strategic sense to maximise national self-sufficiency in the production of weapons, because a long war means heavy demand for the products of the defence-contracting industry, coupled with question marks against the capacity of the defence-contracting industries of allies and others to do much to help - they will

---

[80] John Bulloch and Harvey Morris, *Saddam's War*, Faber and Faber, London, 1991, p. 135.

already have their hands full.

A striking if somewhat homely illustration of the alliance between the navy and the development-industrial level at work and its degree of integration is the so-called Floater. Writing in 1986, Mr J. McGowan - who was shortly to become Director of Marketing at Plessey Defence Systems Ltd - tells us:

> Specific overseas sales [of armaments] efforts in key territories are organised by DESO [the Defence Export Sales Organisation of the Ministry of Defence], often around visits by Royal Navy task forces or the movement of specially equipped Royal Fleet Auxiliary vessels ('Floaters').[81]

And writing in 1978, the then head of the defence sales organisation (the predecessor of the DESO) - Sir Ronald Ellis - indicates that:

> ... we shall be making another sales tour with a floating exhibition [of arms for sale] in the autumn of 1978, visiting European countries and some countries in South America. (The last 'floater' went out in 1977 to the Far East.)[82]

Conclusion

The price of ambiguity

So far, the drifting out of line of the operational and declaratory levels of British defence policy may not appear to have been too serious in its consequences. The possible advantages of the resultant ambiguity - very doubtful with regard to the NATO alliance, but more plausible in the case of more shadowy alliances with the publicity-shy traditionalist rulers of the Arabian Peninsula - have been pointed out. But the war in the Falklands draws attention - to put it mildly - to the disadvantages of allowing declaratory and operational levels to drift so far apart. The war occurred partly because Argentina was unaware that Britain still

---

[81] J. McGowan, 'The Future for UK Defence Exports', *Seaford House Papers 1986*, edited by Giles Binney, Royal College of Defence Studies, London.

[82] Sir Ronald Ellis, 'Defence Sales', *RUSI Journal*, Vol. 124, No. 2, June 1979.

possessed a navy capable of retaking the islands. The Argentines can scarcely be blamed for this because in declaratory terms Britain did not possess such a navy - even the Royal Navy had been pretending that it did not possess the appropriate ships - and it is declaratory-level signals that matter when aiming for non-nuclear deterrence. With conventional forces, peacetime operational level signals cannot be picked up by satellite photography, except in an extremely 'noise-prone' way, or even from *Jane's All the World's Ships*.

An unnecessary war is a large entry on the debit side of any ledger. The price in pounds sterling of the failure to bring the operational level into line with the declaratory is harder to assess. In the sense of the overall net balance of national accounts, there may well have been nothing to pay and conceivably some gains, even after allowing for the very considerable financial costs of the Falklands war and its aftermath, because of the off-setting contribution the out-of-area navy and the global role have made to the invisible (chiefly) trade balance. But it seems certain that the British defence budget as such was, and remained, swollen by the distortion created by the maintenance of a global navy. On the other hand, it is difficult to estimate even tentatively the amounts that might have been involved. If the navy had been cut and reshaped as Nott had wished, the naval budget may have been cut by one-third, leading (other things remaining unchanged) to a one-half point drop in the percentage of GDP spent on defence by Britain (e.g. from 4.5 per cent to 4 per cent). But if the ammunition and spares situation of the other two services (mainly) had been improved, as Nott also wished, net savings on the defence budget would plainly have been smaller.

Saying and doing

The final conclusion of this section is that the problems facing senior governmental personnel within the declaratory circle of actors when undertaking a defence review are perhaps more complex than might appear at first sight. It may be that the drifting out of line that needs remedying is sometimes too great to be accomplished at a single stroke, but that in consequently trying to establish priorities and not take on everything at once, new problems present themselves. We have already seen how the British government's possibly over-emphatic declaratory policies on *nuclear* weapons at the time of the Nott review had the effect of weakening the credibility of a review which had chosen to focus on ways of increasing the effectiveness of the contribution made by British *conventional* forces to European defence. And perhaps in a similar spirit of not wishing to take on too much at once, the development-industrial level was dealt with somewhat superficially in the review, almost as if it had no connection with the operational level. An attempt of sorts to bring the development-industrial level into line was begun, but this occurred outside the text of the review in the form of subsequent pronouncements carried in statements on the defence estimates and in utterances of Ministers in parliament and elsewhere. In fact, the review itself specifically gave the development-industrial circle encouragement to pursue policies of exporting arms out of area. This would have the effect, if the analysis presented here is correct, not of depriving the operational circle of a global role, but the reverse, making the job of bringing the operational level into line with the declaratory harder rather than easier. The development-industrial level and its circle are the subject of the next section.

Table 2.1
Oil Production: thousand barrels/day

	1973	1979	1985	1986	1987
OPEC					
Saudi	7690	9840	3750	5540	4650
UAE	1530	1850	1350	1540	1700
Kuwait	3090	2630	1070	1440	1390
Qatar	570	510	340	380	330
Non-OPEC					
Britain	10	1600	2720	2720	2660
Oman	290	290	500	560	580
Brunei	-	230	-	-	150*

1985 Oil Revenue

	Population	Revenue ($US billion)	Rev./head ($US thou)
Saudi	11.5	27.5	2.4
Kuwait	1.7	9.7	5.7
Qatar	0.3	3.2	10.7
UAE	1.3	12.5	9.6
Iran	45	12.9	0.3
Iraq	16	12.5	0.8
Oman (est)	0.7	4.8	6.9
Brunei (est)	0.2	2.4	12

est : author's estimate
- : no figures available
* : 1988

Sources: Geoffrey Heal and Graciela Chichilnisky, *Oil and the International Economy*, Clarendon Press, Oxford, 1991, pp. 44, 45 and 50.
Brunei: The Land and its People, Brunei Shell Petroleum Ltd, Brunei, 1978.
The Economist Atlas, Economist Books/Hutchinson, London, 1989.
ITN Factbook, Michael O'Mara Books Ltd, London, 1990.

Section Three: The Defence-Contracting Industries

For the entire period of our study, declaratory policy in the matter of acquiring conventional armaments for the British armed forces had three specific aims, though these were not necessarily always given an equal weight: to maintain the British defence industrial base; to obtain value for money in the purchase of armaments for the armed forces; and vigorously to pursue joint programmes for the procurement of armaments with European allies.

Inconsistencies and cost

Three charges

The first apparent inconsistency is between one level of defence policy and another. It concerns the maintenance of a defence industrial base on the one hand, and on the other the operational state of unreadiness for a prolonged war on the part of the army and the air force. Regarding a defence industrial base - which we take here, quite uncontroversially, to mean a national industrial capacity to manufacture the widest possible range of conventional weaponry - as a strategic asset in its own right makes sense only in the context of a prolonged general war. Any dependence upon imports for significant items of military equipment then becomes strategically damaging, for during a war of this kind the countries contracted to supply the imports are themselves unlikely to have warlike items in surplus for export purposes. During the period prior to the Nott review, neither the army nor the RAF was operationally prepared for such a war (nor did they become so after the review). Such a war was not impossible even so, but it was an unlikely eventuality, and even the navy's apparent readiness for it, as we have seen in the preceding Section, is open to more than one interpretation.

The second charge also relates to inconsistency between one level of policy and another. At the declaratory level, up to 95 per cent of British defence policy was NATO-orientated; in line with this, promotion of cooperation with European allies in the procurement of weapons was to be given priority. At the industrial level, by contrast, 90 to 95 per cent of all orders for conventional armaments were placed domestically with British companies. This figure remained essentially unchanged between 1970 and 1990.

The third charge of inconsistency is addressed to the declaratory level: maintaining the defence industrial base and getting value for money may obviously not be compatible objectives.

Cost as the fulcrum

Cost indeed is the fulcrum upon which all charges of inconsistency must ultimately pivot. If the British development-industrial circle was the most economically efficient within NATO, there would obviously be no inconsistency between maintaining the defence industrial base (in other words having the armed services 'buy British') and achieving value for money in defence procurement. If it was not, then the declaratory aspects of policy are themselves inconsistent. Similarly, if no premium was being paid (in the form of foregoing cheaper items of military equipment imported from allies) for maintaining the defence industrial base, the strategic insurance it provided - even against a very unlikely contingency - would have been worth having. But if a premium was being paid, it would not have to be very large to become disproportionate to the risk it was intended to cover: a planner's first duty is to prepare for the *most* likely eventualities. Finally, if buying British always, or even nearly always, represented the best value for money in defence procurement, there would be no inconsistency between the declaratory strategy of orientating 95 per cent of defence policy towards NATO and procuring virtually all defence equipment domestically. True, the most probable general war would be a short one, lasting days or weeks, and the strategic argument against importing weapons from allies when preparing for a 'come as you are war' is much weakened. But if British weapons were the best value for money within NATO, importing weapons would itself incur a premium, leading to a smaller defence capability being put at NATO's disposal than might otherwise have been the case. Any complaints, therefore, should be directed elsewhere in the Alliance and not at Britain. On the other hand, if the British defence industrial base did not provide the best value for money in armaments available from within the Alliance, persisting with a buy British policy puts the development-industrial level of defence policy at an angle to the declaratory. The latter is collective and interdependent, the former nationalistic and self-sufficient.

The first two Parts of this Section aim to show that in fact the British development-industrial circle has not always been the most efficient producer of weapons within NATO, and that in some categories of weapons it is strikingly inefficient. The inconsistencies within and between the different levels of defence policy are accordingly real and need to be accounted for. That task is left to the third Part.

I

Stated policy on procurement

We begin in the usual way by tracking from official publications the evolution of declaratory policy on the purchase or 'procurement' of defence equipment. This runs in two halves, divided by a just discernible fault line in 1981, which is interpreted here as evidence of a rather uncoordinated attempt, begun during the Nott incumbency at the Ministry of Defence, to do something about the inconsistencies. Whilst the eruption of the Falklands War was even less relevant to this undertaking than it was to Nott's chief objective of bringing the declaratory and operational levels better into line, it was attended by no greater success. Indeed it may have had even less success.

According to the 1969 Statement on the Defence Estimates, the government wanted to economise on the use of skilled civilian manpower by:

> securing maximum value for money [in the purchase of military equipment for the armed services]. We are also enlisting the cooperation of industry in the greater use of partnership arrangements for equipment that is likely to sell well abroad. In return for *an assurance of firm orders* for the UK forces and in the expectation of profitable export orders, firms are invited to share the cost of development with the government ... We look forward to wider collaboration with our allies in the development of major items of defence equipment.[1] (emphasis added)

According to the 1972 Statement:

> Special attention is being paid to the need to maintain an adequate industrial base for the future procurement programme and to strengthen relationships with Defence suppliers.[2]

The 1976 Statement spoke at greater length on the subject:

> In decisions on procurement, the first consideration is to meet the Services' essential requirements economically and at the right time. Nevertheless, other factors have to be taken fully into account, such as the balance of payments, employment, the preservation of adequate design and production capabilities, and the military need to improve standardisation of equipment with our Allies. For this last reason, and also in the interests of greater efficiency and economy in the use of the defence resources of NATO, it is important to avoid unnecessary duplication of projects in the Alliance. The right solution will more and more frequently involve some form of international cooperation, whether joint development, cooperation in production, manufacture under licence, or straightforward purchase ... When purchase from abroad is the chosen method of procurement, it should not be a

---

1 *Statement on the Defence Estimates 1969*, Feb., Cmnd 3927, HMSO, London.

2 *Statement on the Defence Estimates 1972*, Feb., Cmnd 4891, HMSO, London.

one-way business ...[3]

The new Conservative administration did not, or did not at first, take a noticeably different line from its Labour predecessors. According to the 1980 Statement on the Defence Estimates:

> The defence equipment programme sustains about 200,000 job opportunities within the major defence industries, and about the same number again ... elsewhere in industry. By providing most of the equipment the Services need from our national resources and through collaboration [with European allies], we keep to a minimum the proportion we buy from abroad. Sales of British defence equipment abroad contribute to the balance of payments and generate further returns for the nation on the capital invested in R & D.
>
> The ability to develop and produce arms is thus an important national asset. It ensures supply; it enables British Services' requirements to be met in an appropriate and timely way; it provides domestic employment; it can be paid for in our own currency; it enables us to collaborate, where that is our preferred course; and it offers the prospect of securing foreign exchange through sales. The Government intends therefore to give full support to British industry in providing these benefits.[4]

Nott's ranging shots

The following year saw a sudden slight shift in declaratory policy as far as the procurement of weapons was concerned. As already noted, the 1981 Statement rather than the Nott review was the chosen vehicle. The wording was extremely cautious and diplomatic towards the development-industrial circle though beginning sternly enough by saying that:

> The procurement budget is provided by Parliament for the research, development and production of equipment for the Armed

---

3 *Statement on the Defence Estimates* March 1976, Cmnd 6432, HMSO, London.

4 *Statement on the Defence Estimates* April 1980, Cmnd 7826-I, HMSO, London.

> Forces. It cannot be used specifically to support industry in its civil markets. The defence industrial base was created to satisfy the equipment needs of the Services and not vice versa.

It then adds more diplomatically:

> There are clearly important reasons why we should normally aim to buy British. The arguments for doing so are not limited to the direct benefits to employment and the saving on imports. The long-term commitment of British industry to meeting our defence needs swells the national investment in R & D and technological progress ... It bolsters our defence sales potential and improves British industries' credentials for winning a worthwhile ... part in international collaborative projects. Besides, self-reliance in defence ... has a value in its own right. It may be false economy to opt for a cheaper foreign product if the result is to weaken or lose altogether British industry's own producing capacity in that line.

Plain speaking soon returns with:

> This does not mean we should pay an unlimited premium for a Made in Britain label ... With current financial constraints ... it becomes essential not just to obtain value for money but to cut costs whenever possible. There are many ways to do this, [including] the benefits that can be obtained from better competition and larger production runs supported either by sales prospects or the various forms of collaboration ... The Government is determined to play a leading role in encouraging allied cooperation of this kind [collaborative production].

The piece concludes by blowing both hot and cold in the same paragraph:

> [Within overall procurement strategy] one prime factor is the military requirement. Another is the finite share of resources available. A third is value for money. Within the framework these establish the Government will use the purchasing power of the defence budget to give maximum practicable support to British

> industry - which does not rule out judicious overseas purchase. The precise content of that support is bound to evolve with time, as will the defence industrial base itself.[5]

The final observation on the defence industrial base is particularly ambiguous, it being unclear whether it refers to a possible shrinkage in the defence industrial base (as a result of 'judicious overseas purchase') or its progressive redefinition on a Europe-wide basis as a result of increased allied cooperation in the development and production of weapons. As we shall see, there is evidence that both possibilities were alive in official thinking - one, plainly, does not preclude the other.

The spoken word

That it was intended that something should 'give' with regard to the defence industrial base was made clear in November of the same year (1981) in Nott's evidence to the House of Commons Defence Committee when he made the passing remark that 'we cannot sustain a defence industrial base for its own sake'.[6] Five days later, Nott's speech to the International Institute for Strategic Studies was even more emphatic:

> we have passed beyond the point where we can afford a defence industrial base for its own sake - for its employment benefits alone ... Our defence industries must reflect the constraints and needs of our front line forces, we cannot sustain them as a priority to the requirements of defence itself.[7]

Even so, caution was to be the hallmark. A passage of interrogation before the House of Commons Defence Committee reveals Nott almost in Jekyll and Hyde mood:

> Question (Mr Conlan): ... Is there a conscious decision to

---

5 *Statement on the Defence Estimates 1981*, April, Cmnd 8212-I, HMSO, London.

6 House of Commons Defence Committee, 'Ministry of Defence Organisation and Procurement', Minutes of Evidence, 11 November 1981, HCP 22-i, HMSO, London, 1981, p. 19.

7 *Survival*, Vol. XXIV, No. 2, March/April 1982, IISS, London, contains extracts from Nott's speech, pp. 89, 92.

purchase British where British equipment is available?
Mr Nott: ... 90 per cent of everything we procure is procured from British industry or (*sic*) our collaborative partners and that is on the basis of a conscious decision. Indeed the industrial implications flowing from our decisions are very much in our minds always. Indeed it is quite often the case that it would be cheaper to buy off the end of a long production run from overseas than to buy from British industry on what is by necessity a shorter production run. The only way we can resolve this difficulty of short production runs ... is by selling more of our equipment abroad ...
Mr Conlan: I suppose it can be claimed that because we spend 90 per cent in this country this is a very high figure, but of course the balance of 10 per cent represents in money terms a very large amount also.
Mr Nott: If we were not to go for value for money in British industry and were [e.g.] to spend £100 million more on an item of British equipment than we needed to spend on the purchase of an equivalent foreign product, it only means that that is £100 million less to spend on some other piece of British equipment which might be more efficiently produced. So there is really nothing to be gained for British industry in buying products which are ... uncompetitive.[8]

Some shrinkage of the defence industrial base seems to be implicit in a Parliamentary Written Answer given by John Lee, the then Parliamentary Under Secretary for Defence Procurement, in January 1984, to a question from Mr Kenneth Warren, a Conservative backbencher:

> the government consider that the acceptance of United Kingdom tenders which are internationally uncompetitive does not lead to a sustainable increase in employment in the economy as a whole. It is accepted that there may be short term employment benefits in particular sectors of the defence industry, but the adverse effects of uncompetitive purchases on the economy ... lead to increases

[8] House of Commons Defence Committee, 'Ministry of Defence Organisation and Procurement', pp. 32, 33.

> in unemployment in other areas which negate the assumed flow-back benefits [of buying British].[9]

But typical of the hesitation or caution that characterised government's attempts from the earliest Nott days to bring the development-industrial circle into line, Mr Lee rather spoiled the effect by adding a caveat:

> Defence procurement decisions are nevertheless taken in the light of the associated defence technological, industrial and other factors.

The emerging declaratory policy on defence procurement, if the caveats are ignored, was certainly of a piece with emerging government thinking on economic policy. A senior Treasury civil servant, I. P. Wilson (to whose Seaford House Paper the author owes a considerable debt for directing his attention to matters that might otherwise have been overlooked), makes this point well by quoting from a 1984 Treasury document for internal Whitehall circulation. The document's aim was to give guidance to government departments on purchasing policy, and says, in part:

> if the UK declines to buy foreign goods when they are cheaper than the domestic alternative, the nation will usually get less value for money. The acceptance of United Kingdom tenders which are internationally uncompetitive does not lead to a sustainable increase in employment in the economy as a whole. There may be employment benefits in particular sectors of industry, but these are negated by the adverse effects of uncompetitive purchases on the economy ... [which lead to] increases in unemployment in other areas.[10]

The civil servants at the Ministry of Defence responsible for drafting Mr Lee's Parliamentary reply were plainly very familiar with this

---

9 *Official Report, House of Commons Parliamentary Debates (Hansard)*, Vol. 52, 1983-84, January 16 - January 27, HMSO, London, Col 205w.

10 I. P. Wilson, 'British Defence Programme: Some Personal Observations on Equipment Procurement Policy, Practice and Mythology', *Seaford House Papers 1985* (edited by Giles Binney), Royal College of Defence Studies, London, p. 100.

Treasury document, or a version of it, and were able to negate any adverse feelings even about borrowing its vocabulary.

European dimension

Declaratory enthusiasm for a possible redefinition of the defence industrial base on a Europe-wide basis reached a peak in the 1986 Statement on the Defence Estimates:

> We have stressed ... that the NATO countries ... must share the costs of modern equipment through effective collaboration. The thrust towards European collaboration stems from a number of factors, including pressure from the United States for Europe to make a greater, and more coordinated, contribution to the common defence; and a growing European recognition that a stronger and more cohesive effort within Europe will strengthen the Alliance as a whole ... while at the same time enhancing the strength of the European defence industrial base.[11]

Even this declaration seems to have represented a retreat from the sorts of redefinition of the defence industrial base that had been a part of Nott's thinking at the time of his review. Nott's speech to the IISS in November 1981 had sent up a trial balloon in suggesting that Britain ought to specialise more narrowly in defence production:

> [Britain] could take a unilateral decision to reduce and concentrate its industrial/technological base upon a limited range of weapon technologies. They would probably be high value ones where we were already established, such as electronics, guided weapons and nuclear. Or we could agree collectively within NATO that individual countries or groups of countries should concentrate on developing and producing particular types of weapons thus avoiding wasteful duplication.[12]

In 1984 the Independent European Programme Group - originally set up in 1976 to promote cooperation in the development and manufacture

---

11 *Statement on the Defence Estimates 1986*, I, Cmnd 9763-I, London, HMSO.

12 *Survival*, Vol. XXIV, No. 2, March/April 1982, p. 91.

of armaments among the European NATO members, including France, but which had languished simply because there was very little enthusiasm for such cooperation and an anxiety that if there was, this might alienate the Americans - was given a shot in the arm. This came in the form of new British enthusiasm for which the new Defence Secretary Michael Heseltine is normally given credit. Amongst his aims, according to Martyn Bittleston, writing in early 1990 as an IISS Research Associate whilst on leave from the Ministry of Defence where his last posting had been Assistant Secretary Air (Procurement Executive) 2,[13] was to have the IEPG promote the rationalisation of European defence industries. One consequence of the new vigour of the IEPG was its Action Plan of 1988. According to Bittleston, the Action Plan at least establishes the principle of intra-European NATO competition in the procurement of weapons. However, as Bittleston also implies, the Action Plan as a whole is not a very pro-competition document. Whilst the British development-industrial circle view of genuine international competition is a dim one - as we shall see - its opposite numbers in most other European NATO countries are no more enthusiastic. The Action Plan in particular reflects this in its acknowledgement of the continued importance, albeit in slightly less protectionist terms than in the past, of the European collaborative principle of *juste retour*. (*Juste retour* or 'fair shares' requires that collaborating countries' domestic defence industries should be accorded a share in the development and the production of the item of equipment concerned in strict proportion to the number of units of the item they have contracted to purchase for their armed forces. A definition of *juste retour* permits a definition of the term 'collaboration', when applied to the development and production of weapons jointly between allies. Collaboration is cooperation within the limits set by *juste retour*.)

Industrial reaction

The cautious and diplomatic assault on the development-industrial circle and the sanctity of the 'buy British' principle, begun in the 1981 Statement, nonetheless excited immediate and sustained salvoes of

---

[13] Martyn Bittleston, 'Cooperation or Competition? Defence Procurement Options for the 1990s', *Adelphi Papers* 250, Brassey's for IISS, London, 1990, p. 19.

protest from spokesmen for the management side of the defence-contracting industry. This in itself perhaps explains why caution in the first place was deemed prudent. Extracts from a letter from the Chairman of Marconi Radar Systems Ltd, Sir Philip Watson, to the Member of Parliament Kenneth Warren, dated 12 October 1981, were published along with the minutes of the House of Commons Defence Committee, before which Mr Warren - whom we have already encountered - was appearing as a witness. They read in part:

> I ... have been unhappy for many years about the persistent purchasing philosophy of the MoD in turning too readily to overseas equipment and systems when we need the experience of developing work to keep our industries alive and well. Whilst I was in the navy I was not able to voice my opinions about this of course, but during the seven years I was Director General (Weapons) for the Royal Navy, I pressed as hard as I could to support the policy of using our own industry wherever possible ... I can assure you that I miss no opportunity to impress upon my ex-colleagues, who are still serving, that they must buy British.[14]

There was another letter to Mr Warren, dated 13 November 1981, from the Director of the Society of British Aerospace Companies, Sir Charles Pringle. Sir Charles had retired from the RAF in 1976 as an Air Marshal and had in the same year gone to work for the defence-contracting industry, joining Rolls Royce and Hunting Engineering. His letter went further, claiming calculations done by Smiths Industries, a defence contractor, had shown that even if a British-made item of military equipment was as much as 45 per cent more expensive than the equivalent item from an overseas supplier, it was still to 'the overall advantage of the British economy to buy British'.

The pressure was kept up. Major General Ken Perkins, whom we last encountered in Oman, had retired from the army in 1982 to join British Aerospace (BAe). In his capacity as spokesman for that company he produced an article for the *RUSI Journal* in March 1984 whose first

---

[14] House of Commons. Second Report from the Defence Committee, Session 1981-82, 'Ministry of Defence Organisation and Procurement', Vol. II, HCP 22-II, London, HMSO, pp. 279, 280.

substantive section was headed 'Why Buy British?'[15]

His first reason echoed Sir Charles Pringle's claim about the overall advantages of domestic procurement to the British economy - not, as we have seen, a view shared at the time by the Treasury. But General Perkins had other shots in his locker, citing strategic reasons:

> Independence in the manufacture of weaponry enables independence of military action in an uncertain political climate. The Falklands Campaign illustrated that our national policies cannot rest only on concerted action within the Alliance and should not be exposed to the risk of embargoes by foreign suppliers. Moreover, home based companies will respond more readily to a national crisis than will foreign suppliers. A healthy defence industry also enables the British services to acquire precisely what they need, which would not always be possible if items were bought off the shelf. Furthermore, the problems caused by frequent defence reviews, underspends and sometimes overspends, are more easily handled with indigenous suppliers.

Giving evidence before a House of Lords committee in 1985 Lord Weinstock, the managing director of GEC, a large electronics and electrical firm with, as we have seen, an involvement in defence business amounting to about 20 per cent of turnover,[16] quoted with evident approval a 1911 minute written by Winston Churchill, as Home Secretary:

> I think it is very undesirable to give Government patronage to foreigners. Very strong and special reasons would be required to convince me of the propriety of giving an order to a foreign firm. Of course when the question arises quite nakedly whether you should pay more to a British firm than you would have to pay to a foreign firm for exactly the same quality goods, a very awkward dilemma is created. The good administration of a Department ... ought in nearly every case to prevent such an unfortunate

---

15 Major General Ken Perkins, 'Gentlemen, Players and Referee: A Partnership between Government and Industry', *Journal of the RUSI*, Vol. 129, No. 1, March 1984, pp. 36, 39.

16 In 1991. *The Independent*, 4 Oct. 1991, p. 23

> conjuncture being created. These little stupid attempts to save two-pence tend to bring the whole structure of the Free Trade argument into discredit with the working class electorate...[17]

As late as 1988, the development-industrial circle was still nagging at what it saw as dangerously heretical and non-Churchillian tendencies in the Ministry of Defence. Lord Prior, who left the Conservative cabinet in 1984 to become chairman of GEC, complained to the RUSI that he did not believe:

> that there is significant recognition of British defence industry's capabilities and strengths, nor sufficient desire to utilise them and enhance them. We are the only country in the world which adopts such an equivocal attitude towards our defence industries, sometimes to the extent of attempting to penalise them against foreign competition (for example, whereas at first, the British bid on a particular air-to-ground missile had VAT added to it, VAT was omitted from the foreign bid for the same missile because it was not relevant).[18]

In the same year and addressing the same audience, Air Marshal Sir Donald Hall, who had retired from the RAF as Deputy Chief of the Defence Staff (Systems) to become chairman of Marconi Defence Systems, compared British procurement policy unfavourably with that of France, claiming (on what evidence it is unclear) that what had once been an 'implicit commitment to national procurement' (i.e. to buy British) in Britain had now gone to the wall and that:

> the British defence market has been opened to the widest international competition. And there is no stated policy for the future of the defence industry. No one can dispute the right of any government to seek better value for money in arms procurement. But a preoccupation to obtain the cheapest price while maintaining indifference to the consequences seems to

---

17 *Report from the Select Committee on Overseas Trade*, House of Lords Paper 238-II, Vol. II Oral Evidence, House of Lords Session 1984-85, London, HMSO, p. 474.

18 Lord Prior, 'New Technology and Future Defence Needs: The Industrial Perspective', *Journal of the RUSI*, Vol. 133, No. 3, Autumn 1988, p. 20.

> represent a triumph of dogma over intellect. The British Government is close to 'knowing the price of everything and the value of nothing'.[19]

The Westland affair

By 1988, in fact, declaratory policy had become even more conciliatory towards the development-industrial circle, Mr Lee's parliamentary answer having represented the high-water mark of confrontation. It may be that the alarm within the development-industrial circle would have subsided sooner had it not been for the Westland affair.

By the end of 1985 it had become apparent that the Westland helicopter company was failing. The then Defence Secretary Michael Heseltine wished the company to be taken over by a consortium of European and British firms and to remain as part of the British, or at least the European, defence industrial base. The Prime Minister, Margaret Thatcher, together with other senior members of the declaratory circle, on the contrary favoured more of a 'market' solution. This would involve the company being taken over by the rival and more efficient United States' helicopter manufacturer, Sikorsky. Whether as a result Westland would or would not become a mere producer of Sikorsky designs under licence was not apparently held to be of any great consequence by the Prime Minister's faction. The fierceness of the subsequent battle was such as to lead to the resignation not only of Mr Heseltine, but also of the Secretary of State for Trade and Industry, Leon Brittan. At one point the survival of the government itself was in doubt.

Westland was sold to Sikorsky but the political price paid was high enough to ensure that no future attempts were made to bring the development-industrial circle into line: they had lost a battle but, perhaps without immediately knowing it, they had won the war.

Phony war

What had started as an admittedly cautious and hesitant attempt begun during the Nott period to challenge the sanctity of the defence industrial

---

19 Air Marshal Sir Donald Hall, 'European Cooperation in Armaments Research and Development', *Journal of the RUSI*, Vol. 133, No. 2, Summer 1988, p. 55.

base, degenerated into a phony war. The watchword was still competition, but competition with gloves on. It was competition taken seriously to the extent that the old system of 'cost plus contracts' between the Ministry of Defence and defence contractors was displaced by fixed price arrangements. These in turn encouraged defence contractors to subcontract more work than formerly (in the process becoming themselves 'lead' or 'main' or 'prime' contractors), and it naturally paid them to subcontract on a competitive basis. The old cost plus contracts, by contrast, had encouraged main contractors to do as much as possible 'in house' so as to maximise what they could charge to the Ministry of Defence as overheads.

But in terms of competition it was not taken seriously enough, before Nott or after, for the Ministry of Defence actually to purchase any important item of defence equipment from a foreign supplier in preference to what was being offered by a domestic manufacturer. There were threats to do so in the 1980s, and even apparently narrow squeaks when it looked at times as though a major contract would go overseas, but it never actually happened. The 'distress' purchase of seven Boeing E-3D AWAC aircraft in 1987 from the United States is an exceptional case. This purchase arose from the failure of a British electronics firm (Marconi/GEC) to come within shouting distance of completing a contract awarded nine years previously for the adaptation of Nimrod maritime patrol aircraft to an airborne early warning role. In spite of £1 billion already spent on the programme, it was four years behind schedule when cancelled in 1986.

Even the repeated declarations that collaborative production of defence equipment together with European allies was going to be pursued, let alone be given a competitive edge - as we see below - amounted to nothing at the development-industrial level. By the end of the 1980s, just as much, or as little, of British military procurement came from collaborative projects with European allies as had been the case at the start of the 1970s. In fairness, however, it should be pointed out that the distance in time between the IEPG Action Plan and the end of the period under study is probably too short for that particular initiative to have worked its way through from the declaratory to the development-industrial level.

Before bringing forward evidence in support of the above claims, it is useful to begin by looking briefly at the case histories of three major items of equipment thrown open to tenders from foreign companies during the 1980s. Whilst their selection has been far from random - one is taken from each of the armed services, and choice has been dictated in part by availability of information - read together with the more statistically based account which follows, it will be seen that the story they tell is a reasonably typical one.

Three cases

Spearfish

M. Asteris, an academic economist, tells the story of the heavyweight torpedo.[20] In 1981, the navy decided it needed a replacement for its Tigerfish torpedo. The main competitors were Marconi with its Spearfish torpedo, and the United States' Gould company with its Mark-48. Collaborative production with one or more NATO allies does not appear to have been considered. The United States' model was cheaper. Asteris says that Gould's bid was £130 million (i.e. almost 25 per cent) below Marconi's,[21] but the order was nonetheless given to the British company. The reasons for this collective Ministerial decision (taken at Cabinet level), according to Asteris, were the prospect of a beneficial impact on unemployment in north-west England; the maintenance of a British capability in this area of defence technology; and a not altogether distinct argument concerning export potential.

Spearfish entered service in 1990. The official estimate of the cost of the programme at 1990/91 prices was £1.3 billion.[22]

ALARM

The House of Commons Defence Committee is our chief source of

---

[20] M. Asteris, 'British Arms Procurement: Protection versus Freer Trade', *Journal of the RUSI*, Vol. 129, No. 1, March 1984, pp. 30, 36.

[21] Daniel Todd, *Defence Industries: A Global Perspective*, Routledge, London, 1988, p. 267, gives the value of the heavyweight torpedo contract as £550 m.

[22] *Statement on the Defence Estimates*, 1991, Vol. 2, July 1991, Cm 1559-II, London, HMSO.

information in the convoluted case of ALARM.[23] ALARM is an Air Launched Anti-Radiation Missile designed to defend aircraft by homing onto ground-based air defence radars of all kinds and destroying them. The competition for the RAF requirement in 1983 for this type of missile (for its Tornado aircraft) was between British Aerospace, the lead contractor for ALARM, and the United States' company Texas Instruments and its broadly similar but less sophisticated HARM (High-speed Anti-Radiation Missile). A third option involving an element of British cooperation in the production of the US HARM (Lucas Industries was the British firm involved) - whose development phase was virtually complete by this time - was also considered. In spite of a perceived urgency attaching to this particular RAF requirement; in spite of an RAF preference for HARM; and in spite of the fact that HARM (bought 'off the shelf') was cheaper,[24] the contract went to British Aerospace.

This was, in accordance with the new philosophy, a fixed-price contract, and BAe had, reasonably enough, subcontracted the production of the rocket motor for the missile to what eventually became Royal Ordnance, again at a fixed price. Development of the motor ran into serious trouble, and BAe was forced by 1988 to turn to a German manufacturer (Bayern Chemie) instead. This gave BAe a contractual claim against Royal Ordnance (which, by this time, had passed into the ownership of BAe). But since the government had indemnified Royal Ordnance against claims of this kind in order to make privatisation of the company easier, the original fixed-price contract between BAe and the Ministry of Defence proved to be worthless. By 1988, when a new fixed-price contract was negotiated between BAe and the Ministry of Defence, the programme was £260 million over budget and several years behind schedule. The missile was rushed into service finally in 1991 at the outbreak of the Gulf war. The cost of the ALARM

---

23 House of Commons Defence Committee, Fifth Report, Session 1987-88, *The Procurement of Major Defence Equipment*, HOC Paper 431, London, HMSO, 1988.

24 Because of development-industrial level secrecy, it is impossible to say by how much exactly. According to evidence before the Defence Committee there was little difference between the new 1988 contract price and what it would have cost to buy HARM at this point. This means that HARM would have originally been cheaper by four or five years of the development cost of ALARM, not counting the £260 m overrun. Nonetheless it is to be borne in mind that ALARM incorporates certain technological improvements on HARM. The *Statement on the Defence Estimates* 1991, Vol. 2, Cm 1559-II, puts the final cost of ALARM as £730 m at 1990 prices.

programme, at 1990/91 prices, is officially estimated at £730 million.[25]

The original 1983 decision to buy ALARM rather than HARM was again taken collectively by Cabinet ministers. Their preference for buying British was based on:

> the opportunity it [ALARM] offered for maintaining and enhancing British technology in an area of growing importance for the development of future weapon systems; the fact that it had more potential for development; and the fact that it would sustain more jobs in the British aerospace industry.[26]

Challenger

The third case concerns the choice of a new main battle tank for the army, even though the final Cabinet-level decision, which was to purchase the Challenger 2 from the British company Vickers, was not taken until 21 June 1991, a little outside the time span of our study.

The army's Chieftain tank, dating from the 1960s, was originally meant to be replaced at the end of the 1980s by the so-called MBT80. Official thinking concerning the new tank seems to have envisaged at least the possibility of collaborative production with a number of allies (the USA, Germany and France were all mentioned), with Britain producing the gun.[27] But the collapse in 1979 of the export order from Iran for a modified version of Chieftain (from Vickers) led to the abandonment of the MBT80 programme, which had not in fact got much beyond preliminary discussions; instead the decision was taken to supply the army with the tanks that had been ordered by the Shah. A large number of these would otherwise have been left on Vickers's hands.

This is a classic instance of what the 1981 Statement on the Defence Estimates may have had in mind when it found it necessary to issue the warning that the defence industrial base was created to satisfy the equipment needs of the services and not vice versa. For we learn from Major General A. R. G. Mullens, speaking in 1988 as Assistant Chief of the Defence Staff, Operational Requirements (Land), that this tank,

---

[25] *Statement on the Defence Estimates*, 1991, Vol. 2.

[26] *The Procurement of Major Defence Equipment*, p. xvi.

[27] *Statement on the Defence Estimates 1979*, Cmnd 7474, London, HMSO.

renamed Challenger, was the same old Chieftain except that it had been re-engined.[28] A Rolls Royce engine replaced one formerly produced by Leyland. General Mullens adds that because the Shah's requirements had been less exacting than those of the British army (presumably because Iran's potential opponents were in a different class), 'we [the army] had to take on something which was only of the order of 60 per cent to 70 per cent of what we wanted'. The Challenger 1 began to enter service in 1983. By the end of the 1980s the army had 450 Challenger 1 tanks which replaced about one-third of the old Chieftains. A long-term plan to replace the remaining Chieftains with MBT 2000, a tank to be collaboratively produced with Germany, was not seen as capable of being implemented soon enough - in view of the age and (as we have seen) the obsolescence of the Chieftains. Thus the decision was taken in 1988 to go for an interim replacement for Chieftain.

There were three overseas 'off the shelf' contenders for the new order: the US Abrams M1A1, the German Leopard, and the French Leclerc. The domestic bidder was Vickers with Challenger 2 - the original Challenger chassis and engine but with a new gun and improved turret. Early Ministerial comment in March 1988 by Lord Trefgarne, who had become Minister of State for Defence Procurement in 1986, might have been seen as settling the issue in advance:

> I and my ministerial colleagues are looking for the most cost-effective solution to the requirement to replace the Chieftain tanks. But naturally we shall take into account the implications of such a decision upon the British tank manufacturing industry.[29]

But in fact suspense seems to have been maintained to the end. Two months before the estimated £300 million order finally went to Vickers (the only British tank manufacturer), the managing director of the company, Sir Colin Chandler, was unable to contain his impatience at the delay in announcing the winner of the competition. He was quoted as accusing the then Defence Secretary, Tom King, of jeopardising Vickers's tank business and of betraying the principles of the

---

28 *The Procurement of Major Defence Equipment*, p. 55.

29 Quoted in *The Procurement of Major Defence Equipment*, p. 54.

Conservative party by not supporting the entrepreneur.[30] The final 1991 order was for only 130 tanks, perhaps in keeping with the radically altered strategic circumstances of the day.

The statistical picture

Equipment for the armed services over the period between 1970 and 1990 has come from three sources: domestic; collaborative production with allies; and direct purchase overseas. Moreover, it is generally agreed that the relative proportions of equipment obtained within these three categories have remained almost constant (by value) over the same period of time, at 75 per cent, 15 per cent and 10 per cent respectively.

One uncertainty in the interpretation of these figures which is comparatively easily disposed of arises from the fact that the development-industrial level is as secretive as the operational level.

Official secrecy means that published figures do not distinguish between purchasing nuclear weapon-related equipment for the armed forces and conventional equipment that is our present concern. As we saw in Section One, nuclear weapon-related equipment normally accounts for only a small proportion of total procurement expenditure in Britain. We also know that about 40 per cent of the cost of the Trident programme is being met in the United States[31] (Trident is, in this, reasonably typical of UK nuclear programmes, although the Chevaline programme, which also falls within our period of study, had a higher domestic content). On this basis we can predict in broad terms the sorts of errors likely to be involved if we nonetheless use aggregated data as a substitute for data on conventional procurement. In fact, the aggregate data will understate the degree of domestic purchase and overstate the extent to which contracts are placed with overseas suppliers for conventional equipment.

The largest and the smallest

With this provision in mind, we can see that by far the largest of the three-part procurement comprises contracts with British industry pure

[30] *Sunday Times*, 21 April 1991, 4.7.

[31] *Statement on the Defence Estimates* 1988, Cm 344-I, HMSO, London.

and simple. This amount has not been absolutely fixed between 1970 and 1990. It accounted for a proportion rising in a shallow parabolic trajectory from about 75 per cent during the 1970s to 80 per cent in the mid-1980s, and falling again gently to 75 per cent by the end of the decade.[32] The third part consists of purchases from overseas manufacturers, falling from about 10 per cent of the total in the 1970s down to around 5 per cent during the mid-1980s, and rising again towards 10 per cent by the end of the 1980s.[33] This trend in overseas purchases (and its mirror image trend for contracts with British industry) is probably partly real and partly, as far as conventional weapons are concerned, spurious (being a reflection of the rise and fall in purchases of nuclear weapon-related equipment from the United States. As we saw in Section One, in the middle of the 1980s procurement of conventional equipment accounted for about 95 per cent of all equipment expenditure, but by the end of the decade, with rising expenditure on the Trident programme, the proportion was down to 85 per cent).

Distributing the middle

The second part is spent in joint programmes with NATO European partners in the collaborative production of defence equipment. This proportion has been even steadier, at 15 per cent - deviating slightly upwards, to 16 per cent, in the second part of the 1980s, according to official figures[34] - although it may be doubted whether the difficulties associated with making such estimates (one important example of which we have referred to above) would entitle us to regard a single percentage point difference as statistically significant.

At first sight, then, it might appear that the charge of inconsistency between the development-industrial level and the declaratory level is misplaced. In the period 1970 to 1990, British declaratory defence policy became increasingly bound up with NATO and collective defence, to the point where 95 per cent of policy was officially declared to be NATO orientated. At the declaratory level, then, defence was almost completely a matter of interdependence with NATO allies: operationally

---

32 *Statement on the Defence Estimates 1973*, Cmnd 5231, HMSO, London and *Statement on the Defence Estimates* 1992, July, Cm 1981, HMSO, London.

33 *Statement on the Defence Estimates* 1992, July, Cm 1981, HMSO, London.

34 *Statement on the Defence Estimates* 1986, Cmnd 9763-I, HMSO, London.

we know this to be something of an exaggeration, though a substantial measure of operational level activity, *pace* the navy, was collective. If the declaratory and operational degrees of cooperation with allies within NATO were to be reflected at the development-industrial level, one would expect the extent of import penetration in terms of defence procurement to be similar to that of manufactured products in general, though perhaps slightly smaller to reflect an insurance premium against the (not very likely) eventuality that a major European war could turn out to be prolonged. In 1972, 17 per cent of domestic demand for manufactured goods of all kinds was satisfied by imports. By 1978, this figure for import penetration for manufactures (i.e. imports *divided by* home demand) had risen to 25 per cent; by 1983 to 31 per cent, and by 1989 to 37 per cent.[35] Similar figures for import penetration in the manufacturing sector, incidentally, apply to France and to West Germany. Some mismatch might be expected in that sources of defence equipment do not perfectly overlap those for manufactured imports in general. Japan, for instance, happens not to be a major independent producer of arms. Nor would other possible sources of manufactured goods be regarded as appropriate suppliers of defence equipment - the former Soviet Union for instance. But since in general most British trade is with NATO partners, any mismatch between the two levels of import penetration is almost certainly not a trick of statistics, but due to real considerations.

For our period of interest, then, import penetration in manufactured goods generally seems to have averaged at about 30 per cent, whilst on a rising trend. Import penetration in defence equipment has been steadier. With domestic production at 75 per cent, it might seem as though import penetration has been about 25 per cent, not really very different from the 30 per cent that might be expected. However, closer examination of the nature of the collaborative projects with European allies shatters this assumption.

Multinational collaboration

It quickly becomes obvious that because collaboration is indeed not the

[35] House of Lords, *Report from the Select Committee on Overseas Trade*, Vol. I - Report, and Vol. II - Oral Evidence, HMSO, London, 1985, pp. 19 and 149, respectively. *Monthly Digest of Statistics*, Central Statistical Office, HMSO, London, June 1992.

same thing as cooperation, the history of British involvement in collaborative procurement programmes has been one of ensuring that the 15 per cent of the procurement budget involved is spent in Britain with British arms manufacturers.

Genuine cooperation at the industrial level with NATO partners could have implied a number of options. At one extreme, it could have meant progress towards a completely common market for armaments of all kinds within NATO, or at least within European NATO (which, after making due allowance for France, is reasonably congruent with the European Community). Less ambitiously, but still rewarding strategically and economically, it could have meant simply a division of labour between European NATO members. This, as we have seen, was one of the kites flown by John Nott in his 1981 speech to the IISS. It would be an echo at the industrial level of what was a fact at the operational and declaratory levels - for instance, West Germany's leading role in the land forces of the alliance and Britain's leading role in its naval forces. Industrial level division of labour could have been made along the cleavage lines between whole categories of weapons, with Britain specialising in air equipment, say, and West Germany specialising in land equipment, or within categories, with Britain specialising in aero-engines and Germany or Italy on air-frames.

Instead of cooperation came collaboration, which can be described as a form of cooperation with allies which is designed neither to promote division of labour nor specialisation through 'playing to strengths', but simply to maintain the domestic defence industries of the participants. This is symbolised by the hallowed NATO collaboration principle of *juste retour*. This, as already noted, requires that collaborating countries' domestic defence industries be accorded a share in the development and the production of the item in strict proportion to the number of units they have contracted to purchase for their armed forces. 'Item' need not necessarily even mean a complete unit: for an aircraft, the engine is normally treated as a separate item from the rest. Even so, collaboration has two positive features which it shares with more genuine forms of cooperation.

In mitigation

One advantage is a tendency to promote commonality in weapon types as between allies, who might otherwise have manufactured or purchased a number of different aircraft, say. This in turn can be expected to have a number of benefits - interchangeability of spares, for instance - at the operational level and in particular on the battlefield. Another is a tendency to reduce the unit cost of an item whenever there are marked economies of scale to be obtained. This is achieved by offering a better prospect (not quite a guarantee) of a larger market for the product than could be obtained if participants in the scheme were acting alone. Each participating country's promise to purchase the product is often seen as a more certain route to long production runs than a small national requirement boosted by success in the export markets. Economies of scale are characteristic of the manufacture of products which incorporate the results of a recent programme of research and development.

Because such products involve a large element of 'novelty', learning how to produce them most efficiently has in some significant measure to take place on the job; thus the first batches are produced less efficiently than succeeding batches which benefit from the so-called 'learning curve' effect. Economies of scale are also characteristic of products where the investment in research and development (R & D, which is made for the most part prior to the production of the first unit) is large - that is to say, comparable to between a quarter to three-quarters of production costs. Defraying R & D costs over as many units as possible is also obviously sensible. But economies of scale do not continue indefinitely. Beyond the production of a few hundred aircraft, for instance, renewal of some of the tools involved in production may be called for and new expense incurred. It is also common for a product like a military aircraft to undergo modification after it has entered production in order to prolong its useful life. Thus, when later batches of an aircraft (which might otherwise have benefited most from the economies of scale) are deemed less up to date than they might be, additional research and development expenditure and modifications to the production process become necessary.

The EFA example

The reduction in unit costs within a collaborative programme is by no means as great as that obtainable within a single, national programme achieving a comparable run of production, partly because of *juste retour*. Thus the original programme for the European Fighter Aircraft (EFA or Eurofighter 2000) had in 1988 four participating NATO members - Britain, Germany, Italy and Spain. Each country had declared an intention to take the following numbers of the aircraft:

Britain:	250 aircraft
West Germany:	250 aircraft
Italy:	165 aircraft
Spain:	100 aircraft
	765[36]

This meant that Spain was assigned 13 per cent (100/765) of the development and production work of the air-frame and 13 per cent of the development and production work of the engine, the two parts of the programme being in the hands of two separate multinational consortia. But to quote Martyn Bittleston:

> Spain's weakness in aero-engine technology means that it will be unable to carry out enough work on this part of the [EFA] project to fulfil even its modest 13 per cent share. But *juste retour* still applies, and the consequence is that all the partners have had to agree to an economically inefficient compensation mechanism to make up the Spanish share.[37]

In general, *juste retour* means that placing work where, even within the collaborating countries, it can be done most cheaply, is ruled out of court. It also means a risk that some of the 'learning curve' economies of scale made possible in principle by collaboration are denied the participants in practice. This could arise whenever *juste retour* requires there to be a series of parallel production lines, none or not all of which have enough throughput of work to generate worthwhile returns to scale. The official Ministry of Defence estimate in 1988 of the savings in unit cost which the collaborative EFA project offered over a purely national, British project to produce the same number of a comparable aircraft for the RAF (i.e. with a production run of 250 opposed to the EFA's planned 765) was 20 per cent.[38] Other informed estimates have put the

---

36 House of Commons Defence Committee, Fifth Report, *The Procurement of Major Defence Equipment*, House of Commons Session 1987-88, HMSO, London, p. viii.

37 Bittleston, *op. cit.*, p. 19.

38 House of Commons Defence Committee, *op. cit.* p. ix.

savings at nearer 5 per cent.[39]

Delay and selectivity

Another cost incurred by collaborative programmes is that likely to be involved in learning how to collaborate. This not only means money costs, but also seems to result in delays which may have damaging operational consequences, even if these are difficult to quantify. According to the House of Commons Defence Committee reporting in 1988:

> Collaborative programmes undoubtedly offer benefits, but also risks, one of which is timescale. We note MoD's [Ministry of Defence] satisfaction with the Tornado programme which it presented as the model of a successful collaborative endeavour. We do not wish to detract from the achievements of the Tornado programme, which has produced highly capable aircraft, but we observe that whilst the ... Tornado prototype and the American F-15 both began their development at the same time, the F-15 was starting a mid-life improvement programme as the Tornado entered active service.[40]

Delay was the reason given by the US government for its withdrawal in 1989 from the embryonic and later aborted NFR90 NATO frigate programme, which had as many as eight collaborating countries involved at one stage.

Another sense in which collaborative projects fall short as substitutes for genuine inter-allied cooperation at the industrial level is that their applicability as far as Britain is concerned seems to be limited. In fact, if not in principle, collaboration is almost entirely restricted to aircraft, or rather aerospace which subsumes aircraft, helicopters and guided missiles. Collaborative programmes for naval shipbuilding have been attempted, or have at least apparently been given serious international consideration (the NATO frigate, NFR90, has already been mentioned), but without success. The same is true of tank production. Thus the Royal Navy's ships have continued to be built exclusively in British

---

[39] Bittleston, *op. cit.*, p. 60.

[40] House of Commons Defence Committee, *op. cit.*, p.ix.

shipyards and the army's tanks in British foundries, even if some of the warlike fittings to these 'weapons-platforms' may have been procured partly or wholly from overseas. It is not surprising, then, to learn from Bittleston that:

> After 25 years or more of collaborative ventures, as of early 1990 the UK has only four out of 39 major projects that are the subject of inter-government collaborative agreements.[41]

Whilst the value or potential value of these four would almost certainly have been higher than average, Bittleston goes on to say that:

> ... 75 per cent of the UK's equipment expenditure is with its national industry. If the work generated by collaborative projects is included, this figure rises to 90 per cent.[42]

At the industrial level, it seems that very little has altered over the timescale of our study, for in 1973 the official figure for the proportion of equipment expenditure placed at home was 91 per cent.[43]

The prize

Of the conventional procurement 'cake', then, British industry - the defence-contracting industry - in one way and another receives between 90 and 95 per cent of the total. During the period of our study the size of this cake increased in an almost perfectly steady upward trend between 1970 and 1985 from 33 per cent of the defence budget in 1970/71 to 50 per cent in 1985/86.[44] By 1990/91 the percentage had fallen back slightly, to 46 per cent. At the same time, for much of the period the defence budget had been increasing in real terms. In round figures, corrected for 1990 prices using the 'all items' retail price

---

41 Bittleston, *op. cit.*, p. 10.

42 *ibid.*

43 *Statement on the Defence Estimates 1973*, Cmnd 5231, HMSO, London.

44 *Statement on the Defence Estimates* 1985, Cmnd 9430-I, and Ministry of Defence, *Defence Statistics*, 1992 Edition, Government Statistical Service, London, 1992.

index,[45] it rose rapidly during the 1970s from £15 billion in 1970 to £21.5 billion in 1980. Between 1980 and 1985 it rose to £23 billion before falling back again to £21 billion by 1990.

Thus the truth of the matter seems to be that, whilst at the declaratory level as much as 95 per cent of British defence policy was on a collective basis with its NATO allies, at the development-industrial level 90 to 95 per cent of procurement was carried out nationally.

Of course there is still the escape hatch of cost. As we have already seen, if the British development-industrial circle could be shown to be the most cost-effective in NATO, all apparent inconsistencies would take on an entirely different complexion.

II

Value for money: the case for

A serious official claim on behalf of the international competitiveness of the British development-industrial circle was advanced in 1988 by Moray Stewart, a senior official (at the time Deputy Under Secretary (Defence Procurement)) in the Ministry of Defence:

> It is true that 75 per cent of our procurement budget goes directly to UK firms, and a further 15 per cent to them through our participation in collaborative projects. But that is overwhelmingly because British companies are able to offer value for money.[46]

Although Stewart's assertion concerning value for money was subsequently described as 'arguably optimistic and certainly unproven' by his Ministry of Defence colleague, Martyn Bittleston,[47] it needs to be considered seriously.

If Stewart is right, then the misalignment between declaratory and development-industrial levels of defence policy becomes moot. If Britain has a comparative advantage in the production of armaments,

---

[45] *Retail Prices 1914-1990*, Central Statistical Office, HMSO, London, 1991.

[46] J. Moray Stewart, 'Defence Procurement in Britain', *RUSI Journal*, Vol. 133, No. 4, Winter 1988.

[47] Bittleston, *op. cit.*, p. 63.

which is what Stewart maintains, cooperation with allies in arms production - in the sense of buying arms from foreign producers on the sort of scale that would lead to an import penetration of, say, thirty per cent - would involve Britain *and* its allies in paying a premium. Fewer or less effective arms would be purchasable for a given outlay and the alliance would be weaker than it is, or the British defence budget larger than it is. As things stand (Stewart might argue), the alliance is weaker than it otherwise might be, not because Britain is pursuing a wilful policy of nationalistic and protectionistic self-sufficiency in the production of armaments, but because Britain's *allies* are doing so, at least to the extent that they are failing to purchase British armaments in the quantities merited by the value for money they represent.

Facts from figures

What evidence is there to support the Stewart line? If Britain does possess a comparative advantage in the production of arms, then we should expect this to be demonstrated in the statistics for international trade. One test is that of 'revealed comparative advantage', employed by the Organisation for Economic Cooperation and Development (OECD).[48] The test can be used to compare the ratio of exports to imports for armaments (for Britain) to the comparable ratio for manufactured goods as a whole. If the former is significantly greater than the latter, this would constitute evidence, at least on the surface, that a comparative advantage exists in the defence sector.

Unfortunately it is easier to state the nature of this particular test than to carry it out. Secrecy at the development-industrial level is again an obstacle, and wholly reliable figures for British exports of armaments are not available. To quote Sir Frank Cooper, writing in 1990, eight years after his retirement as Permanent Under Secretary at the Ministry of Defence, but only a year after leaving the chairman's job at United Scientific Holdings, a defence-contracting company:

> Not a great deal is known about the exports of UK defence equipment because there is no adequate recording system. Reference is placed on partial figures from Customs and Excise and there is little doubt that exports are substantial and, at least on

[48] *OECD Economic Surveys United Kingdom 1988/89*, OECD, Paris, 1989, pp. 28, 35.

> the surface, appear to be ahead of defence imports by about eight times ...[49]

In fact Sir Frank's caution seems to be justified. 'Eight times' would appear to be a considerable exaggeration, and probably arises from an under-recording of imports in some compilations of official statistics. For instance, the table in the official publication *Defence Statistics* dealing with imports of defence equipment is careful to qualify its listings as pertaining to 'identified' defence equipment. Over the period 1985/90, total equipment procurement expenditure at current prices was £45.3 billion.[50] Imports, at 10 per cent, would have amounted to £4.5 billion; at 5 per cent they would have totalled £2.3 billion. Over the same period, total exports of defence equipment, including official estimates of 'additional aerospace equipment', amounted to £12.9 billion.[51] The ratio is then at least 2.9 to 1 in favour of exports. As has already been hinted, the figures used here for British defence exports are themselves not free from controversy. The Stockholm International Peace Research Institute - a body with a history of specialisation in the analysis of the international trade in arms - is unable to reconcile its own consolidated estimates of British arms exports for the period 1979 to 1988 - $16 billion - with the official British governmental consolidated figure over the same period of $50 billion.[52]

Besides being consistent with the £12.9 billion figure quoted above, we have taken the official figure as more likely to be correct for four reasons. First, it corresponds reasonably closely to the sum of the appropriate figures published annually in the statistical annexes to the Statement on the Defence Estimates, provided that so-called 'identified defence equipment' is added to 'estimates of additional aerospace equipment'. Secondly, it squares with semi-independent claims that 'defence exports account for a little over one-quarter [of annual sales of

49 Frank Cooper, 'Economists and Defence: The Views of a UK Practitioner', *Defence Economics*, Vol. 1, p. 81.

50 Ministry of Defence, *Defence Statistics*, 1992 Edition, Government Statistical Service, HMSO, London.

51 *loc. cit.*

52 Stockholm International Peace Research Institute, *SIPRI Yearbook of World Armaments and Disarmament 1990*, OUP, Oxford, 1990, p. 226.

the British defence industry]',[53] and that 'sales to overseas customers account for nearly a third of our national defence production'.[54] Thirdly, it is supported by a published CIA estimate of British arms exports for the years 1987 and 1988 which comes to about $9 billion for these two years alone (albeit valued at 1992 prices).[55] And fourthly, it is particularly hard to see why the British government should wish to overstate its contribution to the international trade in arms to this extraordinary degree.

Taking the export:import ratio as three to one, there certainly seems to be a revealed comparative advantage. For manufactured goods as a whole, the ratio of exports to imports for Britain fell during the 1970s from about 1.6 to 1 to about 1.15 to 1. Since 1980 the ratio has fallen to well below unity, stabilising at the end of the 1980s at around 0.9.[56]

Expert witness

Other evidence in support of the Stewart position is of a more qualitative nature. It is an old saying that if Britain has a particular talent in manufacturing, it is in the manufacture of items where less than mass production is called for. Britain can make racing cars successfully: 90 per cent of all racing cars in the world, from Formula 1 downwards, are made in Britain,[57] but this is small-volume, technology-intensive 'craft' manufacture. The country is much less competitive internationally in the mass production of ordinary motor cars. As Stopford and Turner contended in 1985, British conditions and managerial approaches are most conducive to success when problems of mass scale need not be confronted.[58] The production of modern armaments probably has more in common with that of Formula 1 racing cars than of VW Beetles.

---

53 Bittleston, *op. cit.*, p. 37.

54 *Statement on the Defence Estimates 1986*, I, Cmnd 9763-I, HMSO, London, p. 48.

55 Congressional Budget Office, 'Limiting Conventional Arms Exports to the Middle East', Congress of the United States, Washington DC, September 1992, p. 13.

56 *Monthly Digest of Statistics*, No. 558, June 1992. Other sources broadly concur. A recovery to 0.94 in 1991 is reported by the OECD. See 'Survey' of Britain, in *The Economist*, October 24, 1992.

57 *The Economist*, September 12, 1992, p. 41.

58 John M. Stopford and Louis Turner, *Britain and the Multinationals*, John Wiley and Sons, Chichester, 1985, p. 98.

Domestic demand for modern military aircraft reaches hundreds at most; for Britain, similar less-than-mass scale runs are deemed appropriate for the production of main battle tanks, while warship construction is scarcely a matter of production runs at all.

A further strand of support for Stewart comes from a study by the OECD of the international competitiveness of British industry.[59] This found that during the 1980s comparative advantage was gained by those sectors of industry that were research-intensive. The research-intensiveness of the defence sector of British industry is high, even notoriously so.

What also makes the defence sector of the research-intensive part of British industry unusual is government's role in meeting much of the capital cost represented by research and development. According to a valuable statistical 'X-ray' taken by the Advisory Council on Science and Technology (ACOST) for 1986[60] concerning matters that are normally, perhaps for reasons of commercial confidentiality, left rather obscure in other statistical compilations, the Ministry of Defence provided about 45 per cent of the total R & D funds of the largest defence contractors, rising to over 50 per cent in the case of aerospace and electronics contractors. The average for the defence industry as a whole is much larger, at 83 per cent. This is partly because smaller companies are less capable than large of carrying even a share of the risk represented by R & D investment, and partly, it seems reasonable to assume, because some sectors of the defence-contracting industry are less profitable than aerospace and electronics. According to ACOST, total R & D expenditure by the defence-contracting industry from its own resources in 1986 was roughly £350 million, at current prices (plus or minus £50 million - R & D figures are notoriously imprecise - see the appendix to this Section). According to the appendix, the figure for 1989, again at current prices, was about £400 million, with a combined government and overseas contribution of £1220 million, plus a government intramural spend of £940 million. Depending upon how much of the latter is regarded as R & D that the defence industries would have needed to indulge in anyway to get their products from the research stage to the market place, and recalling that as much as 50 per cent of the £940 million may relate to nuclear weapons, the ratio of R

---

59 *OECD Economic Surveys United Kingdom 1988/89, loc. cit.*

60 ACOST, *Defence R&D: A National Resource*, Cabinet Office, London, HMSO, 1989.

& D to total sales for the defence-contracting industries is between 16 and 25 per cent.

Compared to chemicals and pharmaceuticals, another successful sector of British industry by the criterion of revealed comparative advantage, although with an export to import ratio of around 1.3 in 1986 (and on a slowly declining trend),[61] the ratio of expenditure on R & D to sales in the defence-contracting industry seems very high. For chemicals and pharmaceuticals, where there is little or no government subvention, the ratio is around 5 per cent.[62] Nonetheless ACOST seems to come down, with OECD, on Stewart's side:

> defence contracting in the UK is a successful business. In addition to the MOD's procurement, the industry has earned substantial export earnings ...

A third piece of evidence for the Stewart case also comes from Stopford and Turner. As we saw in Section One, they cited the following ten UK (non-oil) firms in 1985 as possessing 'global stature': Pilkington, BOC, Reuters, Beecham, Racal (since 'demerged' into Racal Electronics and Vodaphone), Plessey, Glaxo, British Aerospace, Rolls Royce (aero-engines) and 'parts of' GEC.[63] Of these ten, we learn from ACOST that five - British Aerospace, GEC, Rolls Royce, Plessey (taken over in 1989 by GEC and Siemens, the German firm) and Racal - were (in 1986), in that order, the leading companies in the country in terms of the portion of their turnover deriving from defence business.

The case against

But the case for the prosecution is at least as formidable.

To begin with, the defence-contracting industry in Britain does not operate in a vacuum. To quote from Bittleston:

> ... there are wide variations in company exposure to defence

---

61 See Report from Select Committee on Overseas Trade, Vol. I - Report, p. 18 and *Monthly Digest of Statistics*, June 1992, CSO, HMSO, London.

62 *The Independent*, 9th June, 1992.

63 Stopford and Turner, *op. cit.*, p. 105.

> spending in both the engineering and electronics sectors, even though, overall, in both groups, defence sales account for 30 per cent of the revenue of major companies. In general, the UK's large defence companies are more dependent on defence ... than their counterparts in the US and the FRG, but less so than the 'private' French aerospace companies. Within the aerospace sector, where BAe, with a turnover of $9.0 bn (civil and military) dominates and Westland and Shorts are major players, defence absorbs about half the output. In shipbuilding, where defence orders are split between the submarine constructors Vickers [Shipbuilding and Engineering], Cammell Laird and Yarrow, the figure is one-third. Estimates in the electronics sector, in which GEC (with sales of $9.3 bn), Plessey, Racal and STC lead, are more variable. Figures range from 15-30 per cent, with official estimates tending towards the higher end of the range.[64]

In the 1970s matters were much the same. According to the Minister of Defence (Roy Mason) in 1974, defence orders accounted for 49 per cent of the output of the aircraft industry and 31 per cent of that of shipbuilding. Only in electronics have things altered. In the early part of the 1970s defence was responsible for only about 10 per cent of the output of the British electronics industry.[65]

Since the British manufacturing sector as a whole shows every symptom of decline since the early 1970s, the relative success of the defence-contracting industry, which as we have seen is interwoven with the manufacturing sector as a whole, is a puzzle. It does not seem to be very plausible to suggest that, within large companies, the defence divisions - irrespective of the ethos, salary scales, traditions and structures of the companies concerned - magnetically attract and retain all the best managers, salesmen, and research and development scientists and engineers. It may be thought more likely that the relative success of the defence-contracting industry in fighting off import penetration and scoring an impressive revealed comparative advantage is due to a deliberate government policy of discouraging imports of armaments. It may even be the case that success in the export markets has not been

---

64 Bittleston, *op. cit.*, p. 39.

65 Roger Facer, 'The Alliance and Europe: Part III Weapons Procurement in Europe - Capabilities and Choices', *Adelphi Paper No. 108*, IISS, London, 1975, p. 14.

obtained on merit either, through the front door so to speak, as a result of the value for money offered by British-made arms, but via the 'back door'. The back door is provided by the global aspect of defence policy and in particular the informal alliance network connecting Britain to the states of the Arab shore of the Persian Gulf (this is not to imply that other important arms-exporters do not use 'back doors' of their own). Statistics in this area must always be treated cautiously, but if official figures for exports of 'identified defence equipment' are a representative sample of all arms exports, 'Middle East and North Africa' accounted for 26 per cent of all British arms exports in 1975; 30 per cent in 1980; 35 per cent in 1985; and 61 per cent in 1990.[66] According to SIPRI, during the 1980s, on average about 30 per cent of all British arms exports of 'major weapons' went to the Gulf theatre of the Middle East, including Jordan.[67] Again according to SIPRI, North Africa (i.e. Algeria) accounted for less than one per cent.

Trade figures

Two further pieces of important statistical evidence need to be introduced before a verdict can be arrived at. The first derives from a previous argument. It is true that the defence-contracting industry is interwoven with British manufacturing industry as a whole, but this statement needs to be qualified. In the first instance, the aerospace sector of the defence-contracting industry is interwoven with the aerospace sector in general; electronics with electronics; shipbuilding with shipbuilding; and so forth. Thus the comparison between the defence-contracting sector and manufacturing industry as a whole is not the only one that can be made. It is useful, in other words, and indeed less crude, to attempt to draw comparisons within industrial classifications. If the civil sector of the classification in question enjoys comparable commercial success to that of the defence sector, this would strongly suggest that the success of the latter had been won on its merits of value for money, on the assumption that the civil sector is exposed

---

66 *Defence Statistics*, p. 17.

67 Ian Anthony (ed.), *Arms Export Regulations*, SIPRI/Oxford University Press, Oxford, 1991, pp. 181, 182. The Gulf theatre is defined as including Saudi Arabia, Iran, Iraq, Kuwait, the United Arab Emirates, plus other member states of the Gulf Cooperation Council.

to market forces, or at least is not under the same degree of suspicion as the military sector that it may not be so exposed.

Electronics

But to begin with, the most important intermediate category of goods bound up with the production of defence equipment is electronics; indeed about one-fifth of procurement expenditure within Britain by the MOD is placed in the electronics sector of British industry, where, as will be recalled, it may account for as much as 30 per cent of production. The electronic products are intermediate in that they may finish up as radars on aircraft and ships; guidance systems and sensors for naval, air and army missiles, as well as for aircraft; or tactical level command, communication, and control equipment within all three services. Electronics can also be an end product: for instance, strategic level command control and communication (including space-based) equipment; training simulators for all three services; or free-standing radars for providing early warning of attack on fixed military installations. The competitiveness of the British electronics industry is therefore a matter of interest. If it is internationally competitive, then the purchase of British electronics for British military equipment should not involve payment of a 'buy British' premium (i.e. the difference between the cost of buying British and the cost of buying the equivalent product from the cheapest overseas supplier).

The revealed comparative advantage test shows an export:import ratio for the entire electronics sector of British industry of 1.1 in 1978, falling to 0.8 in 1985, but stabilising at that level by 1990.[68] Compared to the export:import ratio for the British manufacturing sector as a whole, these figures are slightly low, even if showing signs of relative improvement. They certainly do not suggest that a 'buy British' policy for military electronics is invariably the commercially sensible thing to do, but nor do they indicate that the premiums paid for doing so will be particularly large.

Aerospace

Turning to end products, the picture within aerospace is suggestive of

[68] Annual Abstract of Statistics, 1991, HMSO, London.

a greater correspondence between buying British and obtaining value for money, even if the indications are again a little mixed. The sector as a whole (civil plus defence) enjoyed a revealed comparative advantage in 1975 of nearly 2; in 1980 1.3; in 1985 1.7; and in 1990 1.25. Some caution is necessary in interpreting these figures. Whilst defence business, as we have seen, accounted for about 50 per cent of total output in this sector by the end of the 1980s, earlier, in 1975 and 1980 (when defence probably accounted for about the same percentage of business), it accounted for around 80 per cent of exports. In this period, then, trade in military equipment clearly dominated the export statistics. Only in the later period is there reliable evidence that the sector as a whole became internationally competitive. In 1985 defence accounted for 45 per cent of aerospace exports; in 1989, total aerospace exports were £5 billion (at 1990 prices), with the defence sector accounting for £2.4 billion or 48 per cent.[69]

Shipbuilding

Aerospace, however, turns out not to be typical. The situation with respect to shipbuilding lends no support at all to the Stewart position. The Nott review of 1981 complained that 'it is a decade since we last sold overseas a major new warship of Royal Navy design'. The position has not materially improved since: it is now two decades since Britain last sold a major new warship of Royal Navy design overseas. In 1989, which was reasonably representative of the decade as a whole, warships accounted for one per cent of all British defence exports. The condition of the civil shipbuilding sector is similarly dire. The revealed comparative advantage test puts the export to import ratio for the shipbuilding sector as a whole at the particularly low figure of 0.62 for 1979, improving slightly to 0.75 in 1984.[70] This impression of poor international competitiveness is confirmed by a finding of Guatem Sen, an academic economist. Sen found that in 1978, subsidies to British shipyards amounted to the very high figure of £50 million for each

[69] *Defence Statistics* 1992 Edition and *The Independent on Sunday*, 26 July 1992.

[70] *Input-Output Tables for the United Kingdom 1979*, PA 1004, Business Monitor, Department of Industry Business Statistics Office, HMSO, London, 1983, and *Input-Output Tables for the United Kingdom 1984*, Central Statistical Office, HMSO, London, 1988.

Royal Navy ship constructed[71] - a premium of well over 50 per cent of the cost of a frigate to the taxpayer. Indeed British shipyards seem to be virtually unable to attract orders from anywhere but the Ministry of Defence and for anything but warships. It is said that in 1983 the five pre-privatisation yards of British Shipbuilders then engaged on Admiralty work owed 72 per cent of their turnover to MOD contracts.[72] In 1982 the Controller of the Navy, Vice Admiral Bryson, told the House of Commons Defence Committee that 'the profitability of British Shipbuilders over several years has rested largely on the very solid base of orders which we [the navy] have provided'.[73] (British Shipbuilders was de-nationalised in 1985.) The obvious conclusion to be drawn is that the value for money offered by major British warships is so poor - because British shipbuilding practices are so inefficient - that even the semi-client states of the Arabian peninsula cannot be persuaded to buy British. Correspondingly, when the navy buys British, as it invariably does, a substantial premium must be paid.

Main battle tanks

The situation with respect to armoured fighting vehicles and artillery is better than that of warships, but possibly not by very much. Over the period of our study only one contract for a major sale of main battle tanks overseas has been implemented in full. Jordan purchased over 200 Chieftain tanks in 1979. A very large order for 1500 modified Chieftain tanks to go to Iran collapsed with the fall of the Shah in the same year. Altogether it appears that only about half of the Chieftains were ever delivered to Tehran. As we have already seen, the surplus tanks destined for Iran were renamed Challenger and bought in for the British army.

The official statistics confirm the story of doubtful British competitiveness in the production of army equipment. 1989 was a particularly poor year for the export of army equipment, so that it would be unfair to cite it; but even 1986, which was much more representative,

---

[71] Sen's work is quoted by Miss C. S. Rycroft, 'Self Sufficiency: A Worthwhile Quest in Agriculture and Defence Equipment?', *Seaford House Papers 1984*, (edited by Giles Binney), Royal College of Defence Studies, London, p. 24.

[72] Daniel Todd, *Defence Industries: A Global Perspective*, Routledge, London, 1988, p. 184.

[73] 'Ministry of Defence Organisation and Procurement', p. 293.

saw armoured fighting vehicles and parts account for only 4 per cent of all military exports. Unfortunately, the civil sector of the economy most closely resembling the defence sector where army equipment is concerned is not very readily identified. Though the heavy end of the motor vehicle industry is one approximation to it. In 1970, Britain had an export to import ratio in cars of 4 to 1. In 1978 it was down to 1.3 and by 1983 it was .55. Since then it has stabilised somewhat. The 1990 figure is .45. The deterioration in the situation with regard to lorries has been similar. Yet the story is not all one-sided. Mechanical engineering is another sector sharing some characteristics with the production of army equipment. In 1978 the export:import ratio here was 1.7; by 1983 it was 1.4. This certainly corresponds to a revealed comparative advantage. Even more striking is the picture presented by the manufacture of tractors and agricultural machinery, which might be thought of as having some sort of industrial kinship to tank and field gun manufacture. In 1979 the export:import ratio was just over 2; in 1984 it was 1.4.[74]

Overall, then, the picture for army equipment is mixed and judgement must be tentative, but in view of the particularly large potential export market for tanks globally, especially in the Middle East, Britain's very poor export record is rather damning. For instance (always bearing in mind the imprecision attaching to all figures relating to the arms trade), according to the US Congressional Budget Office,[75] in the five-year period 1985/89, exports of tanks from all sources to 'developing countries' amounted to approximately 7300. If the British share of the developing-country market for tanks equalled its share of world manufacturing exports as a whole (about 7 per cent)[76], of these 7300 tanks, Britain should have been the source of about 500 in this five-year period alone. Instead, Britain sold none. In the whole of the twenty-year period between 1970 and 1989, in fact, it managed to export fewer than 1000.

It seems likely that when the British army buys British, as it invariably does where tanks are concerned, a premium is again being paid.

---

[74] *Input-Output Tables for the United Kingdom 1979*; *Input-Output Tables for the United Kingdom 1984*.

[75] Congressional Budget Office, *op. cit.*, p. 13.

[76] The figure for Britain's share of world exports of merchandise in 1992 was 5.1 per cent. *The Economist*, April 10th, 1993, p. 120.

How others see us

It may be that the pattern of international collaborative ventures for the production of military equipment that Britain has been involved in with its NATO allies is itself evidence that only the aerospace sector of the defence-contracting industry has been internationally competitive. Such an argument would proceed along the following lines.

Defence-contracting firms abroad, or some of them, are on the look-out for commercial advantage. Foreign aerospace companies have recognised the competitiveness of the British aerospace sector and have accordingly been keen to learn something from it. Hence they have been eager to enter into collaborative production agreements. To quote Bittleston once more:

> At the very least, the technological and industrial competence of the weaker partner [in a collaborative project] is brought closer to that of the stronger. Whether this is either an economically or commercially desirable proposition depends on the observer's standpoint.[77]

By contrast, foreign defence-contracting companies in the naval and land equipment sector have appraised the corresponding British companies as uncompetitive and as having nothing of technological or industrial interest to offer. Hence collaborative ventures in these areas never involve British firms, because they are never courted by overseas companies. Nor do they themselves court overseas companies because, unenterprising in all things, they prefer to shelter behind government-inspired protection against imports than to venture abroad in search of more capable partners in collaborative arrangements.

The 'buy British' premium

How much 'over the odds' has been paid by British governments as a premium for pursuing a 'buy British' policy is exceptionally hard to estimate. The most reliable and certainly the most widely quoted work relevant to this topic is that of the academic economist Keith Hartley,

---

77 Bittleston, *op. cit.*, p. 56.

who estimates that, overall, net savings in procurement costs of as much as 20 or 30 per cent could be achieved were NATO to become a free-trade area in the production, buying and selling of arms.[78] Hartley is speaking of NATO members' procurement budgets as a whole; since defence industries in some member states will be more efficient than those in the UK, and some less efficient, it does not follow that the savings identified by Hartley would necessarily benefit the British taxpayer precisely by the amounts suggested. Even so, the figures are not incompatible with that quoted above from Sen which puts the premium for buying RN warships in Britain at over 50 per cent. The relative competitiveness of the aerospace sector of the British defence-contracting industry partly helps to cancel out the uncompetitiveness of the naval and land warfare equipment sectors. To see this, it is necessary to negotiate yet another obstacle erected by official secrecy in order to decide in what proportions procurement expenditure has actually been allocated to land, sea and air equipment. Official figures, as we have seen, do not distinguish between conventional and nuclear expenditure. However, at times when nuclear expenditure can be presumed to be low and hence not inflating the naval heading unduly, the distribution of procurement expenditure has not varied much in the period between 1975 and 1990 from the ratio land:sea:air is as 2:3:4 (in very round figures).[79] It therefore follows that even alarmingly high premiums being paid to buy British ships (and possibly tanks too) may still be compatible with the Hartley figure overall. If we are justified in applying the Hartley figure to the British conventional procurement budget, the premium being paid for buying British is about 10 per cent of the defence budget. If the premium were not there, in other words, a defence budget of 5 per cent of GDP would reduce to one of 4.5 per cent.

In reality, we should not expect the potential savings to be quite so large. Even if there were no problem in interpreting the Hartley figures, they make no allowance for departures from free trade within NATO that might be justified on the sorts of strategic grounds identified earlier. A limited degree of protection for the development-industrial

---

78 Keith Hartley, *NATO Arms Co-operation: A Study in Economics and Politics*, Allen and Unwin, London, 1983, p. 67.

79 *Defence Statistics* 1992 Edition. *Statement on the Defence Estimates 1980*, Vol. II, Cmnd 7826-II, HMSO, London.

circle as a hedge against the possibility of a long war would have been worth paying for. Indeed it would not have been incompatible with the tenor of the Nott review, in its aim to equip the operational level with the means to fight a longer European war than the one of only a few days' duration which the army and the RAF were (and remain) kitted out for. Perhaps for this reason the ten per cent premium being paid for buying British should be regarded as an upper limit.

The verdict

Even so, the conclusion drawn from this test of the validity of the Stewart position is that it is only about half right and that Bittleston's scepticism is justified. There are major and minor consequences flowing from this conclusion.

To take the minor first, since 1983 it has been official policy within the Ministry of Defence to inject competition into the process of defence procurement. As we have seen, during this period no major procurement contracts have been let (except under duress) to overseas suppliers. Even so, a 'competition' policy of a kind may have been in force with regard to the important aerospace sector. The twin facts that this sector is dominated by a single main contractor, British Aerospace, and that there is only one British manufacturer of aero-engines, Rolls Royce, both tend to suggest that the scope for competition in the usual sense of the term must be very limited and that the relative success of this sector may be due to other reasons. One of these is hinted at by Bittleston as likely to emerge from participation in collaborative programmes even when other partners in the programme may not be industrially and technologically on a par:

> it seems reasonable to assume that sharing technical knowledge, and the inevitable rivalry between engineers that goes with this, results in a better product.[80]

Outside aerospace, the failure of British naval shipbuilders to export a single new major vessel in twenty years, and the scarcely more impressive performance of tank manufacturers, would seem to make

---

[80] Bittleston, *op. cit.*, p. 56.

nonsense of any claim that a contract entered into with either of these sectors by the Ministry of Defence can have been on 'competitive' terms. True, whilst there is in any case only one tank manufacturer, Vickers, a handful of shipyards still exist capable of constructing warships (including VSEL with a submarine specialisation, Cammell Laird and Yarrow, the latter owned by GEC). But it would seem to follow from our analysis that neither the one manufacturer of tanks nor any of the shipyards is competitive by international standards.

The major conclusion to be drawn from this survey of the value for money provided by the British development-industrial circle is that the inconsistencies at the declaratory level and between the declaratory and development-industrial levels alleged at the beginning of this Section are real. What now remains to be explained is why they have persisted.

III

Why inconsistencies persist

Some general reasons have already been rehearsed in the previous Section when we considered the persistence of inconsistencies between the declaratory and operational levels. Discontinuity within the topmost layers of the declaratory circle due to changes of government and of ministerial teams will have benefited the operational and development-industrial circles alike. Even the external value of strategic ambiguity, whilst undoubtedly slender, cannot be entirely ignored. An industrial circle seemingly prepared for a long war (like the navy) whilst the army and the air force were prepared for a shorter conflict may have added useful (but transient) extra vagueness to the flexible response strategy.

Moreover, delay in instituting defence reviews and the consequent growth in the scale of the realignment required will have provided mutual protection to the operational and industrial circles. Taking on the navy and the development-industrial circle together would appear to be a tall order. Nott tackled the navy head on, but was able to approach the development-industrial circle only circumspectly. Instead of the broadside of a defence review document, only a few mild sallies were attempted. Even this concentration of fire at the operational level, as we saw in the previous Section, was not fully comprehensive and allowed the nuclear part of defence policy to escape entirely unscathed.

Friends at court

Like the operational circle, the industrial circle too has 'friends at court'. These are typically members of parliament having special connections with the industrial circle. Sometimes this is no more than a matter of representing a constituency where a large proportion of the electorate owe their livelihoods to the defence-contracting industry. Alternatively, there may be a more direct connection in expertise and in sentiment between the member and the industrial circle. To take an example ready to hand, Kenneth Warren, Conservative MP for Hastings and Rye, is an aeronautical engineer by profession whose previous employers had included Smiths Industries. To the extent to which there is an alliance between the operational and development-industrial levels, one can find members of parliament with an armed services background who will also lobby on behalf of the development-industrial circle.

The Treasury official I. P. Wilson even implies that the House of Commons Defence Committee, if not actually a friend at court, is not far from the development-industrial camp. But he advances a particular reason why the development-industrial circle is able to repel attempts to bring it into line. He blames the organisation of the Ministry of Defence and the practices of the Procurement Executive (PE):

> the PE appears to bob uneasily in the stormy water between the rocks of industrial lobbyists and the shoals of stated government policy ... On the one hand it is a purchasing agent for the Services, tasked with providing the weapons and equipment deemed necessary in a cost-effective and efficient way. On the other hand it sees itself as being responsible for the defence industrial sector and preservation of the defence industrial base. These two functions are incompatible and most outside observers would agree that 'industrial policy' is too often allowed to take precedence.[81]

Wilson believes that the Procurement Executive's responsibilities should be pared down to only 'meeting the operational needs [of the armed services] in the most efficient and cost-effective way possible'. Responsibility for the continued protection of the development-industrial

---

[81] Wilson, *op. cit.*, pp. 102, 103.

circle should be passed to the Department of Trade and Industry, which, Wilson claims, is at present in an unholy alliance with the PE and the defence-contracting industry, but without any responsibility for meeting the cost of the hidden subsidies involved in the 'buy British' policy.

NATO allies

Alongside this domestic explanation there is also an Alliance explanation as to why the industrial circle has been so successful in maintaining the size of its defence industrial base. The industrial circles within other NATO members are often just as averse to competition and erosion of their 'defence industrial base' as is their British opposite number. This is not peculiar in any way. Competition may be good for the customer; no one seriously pretends it is good for the company. Because allies also want to preserve their defence industrial bases, the British development-industrial circle benefits from what strength there is in the claim that any British move to buy defence equipment from whichever allied source offered best value for money would not be reciprocated, since allies would continue to favour their home producers.

The obvious retort is that, if true, it matters only inasmuch as it is Britain's allies rather than Britain itself that remain content to receive inadequate value for money from their procurement budgets. But it may not even be entirely true. For one thing, offset arrangements are normally available from the source country. That is to say, customarily when the source country makes its sale, it will additionally promise to spend some proportion of the sum involved back in Britain. Where the gap between promise and reality is not too great, instead of using such funds to keep uncompetitive national defence-contractors going in a lower gear, so to speak, they could be used instead to ease the transitional pain of abandoning uncompetitive industrial defence activities altogether. Secondly, smaller allies do not even attempt to provide their requirements for armaments entirely from national sources. As for larger allies, the United States and France apparently buy an even smaller proportion of their weapons from overseas than does Britain: 2 per cent and 5 per cent respectively (although the high proportion of total spending devoted to nuclear weapons in these countries - virtually all of it being spent domestically - makes international statistical comparisons particularly tricky).[82] In spite of some difficulty in making due allowance for the special historical circumstances under which German purchases of arms abroad were employed to offset the costs in foreign exchange to allied governments of stationing troops in that country, the FRG seems to be less protectionist. There is another kind of uncertainty here regarding definitions, but the FRG probably buys somewhere between 15 and 30 per cent of its armed forces' equipment from overseas.[83]

Defence industrial base

Incidentally, the insinuation of the phrase 'defence industrial base' into the vocabulary of the declaratory circle was itself a minor victory for the development-industrial circle. 'Base' suggests something rather important, not to be lightly tampered with. Chambers Dictionary's principal meaning for the word is 'that on which a thing rests'.[84] This echoes the success of the operational circle in disseminating and insinuating the concept of 'commitment' into the vocabulary of discussion of British defence. However, there are a number of other reasons specific to the development-industrial level which may have accounted for its success in resisting exposure to international competition.

Timing

The first of these is important without being particularly interesting. The tentative moves begun by Nott to question the 'buy British' procurement policy could not, from his perspective, have been more badly timed. In 1983, for the first time, Britain imported more manufactured goods that it exported. The trend towards greater import

---

82 Bittleston, *op. cit.*, p. 10.

83 Bittleston, *op. cit.*, p. 10, gives 30 per cent. On page 34 he says 15 per cent. These figures relate to the end of our period of study. For the beginning of the period, 30 per cent seems to be the correct figure. Roger Facer, 'The Alliance and Europe: Part III Weapons Procurement in Europe - Capabilities and Choices', *Adelphi Papers* No. 108, IISS, London, 1975, p.35. Like Bittleston, Facer wrote his *Adelphi Paper* whilst on leave from the Ministry of Defence.

84 *Chambers Twentieth Century Dictionary*, Chambers, Edinburgh, 1980.

penetration and a decline in employment in manufacturing have continued. This was no time, it will have seemed to many at the declaratory level, to test the strength of the argument of Churchill's 1911 minute that giving 'government patronage to foreigners' in the shape of arms contracts could not be expected to be understood by the electorate. Manufacturing is labour intensive, and government decisions which added in a very direct way to already rising levels of unemployment in the industrial sector, however transitionally, would have taken an unusual degree of statesmanship to implement.

The importance of the protection which this state of affairs gave to the development-industrial circle was as a *deus ex machina* and roughly parallels the protection the Falklands crisis gave to the navy. That it is not as interesting is due to the fact that, whilst accident and chance played a part in both cases, the state of the manufacturing sector of British industry can have owed little or nothing to the activities of the development-industrial circle. The latter was too small. Between 1975 and 1980 manufacturing output in Britain was around 25 per cent of GDP but falling. By 1988 there had been a recovery to 24 per cent of GDP (employment in the sector did not recover; the manufacturing workforce declined by over 2 million between 1979 and 1989). At about fifty per cent of the defence budget, the output (including exports) of the defence sector has never been more than roughly a tenth of all industrial output. And of this tenth, about half - the aerospace part - seems at times to have been at least as productive as manufacturing as a whole - more of a buoyancy ring than a dead weight around the floundering manufacturing sector, one might think. If the 'tails seldom wag dogs' argument does not convince, the lack of any evidence that large defence budgets, swollen by uncompetitive procurement processes or otherwise, automatically have damaging effects on the economy (beyond the opportunity costs of other goods or services foregone) should. According to a survey led by Keith Hartley:

> Economists who have looked for evidence that military expenditure has a unique effect on economic performance have found mixed results.[85]

---

[85] Keith Hartley, Nick Hooper, Steve Martin and John Singleton, 'The Economics of UK Defence Policy in the 1990s', *RUSI Journal*, Vol. 135, No. 2, Summer 1990, pp. 51, 52.

By contrast, the previous Section suggested that the failure to bring the declaratory and operational levels of defence policy into line was connected with the outbreak of the Falklands War (although not, of course, with its timing).

Privatisation to the rescue

Whilst an attack on the 'buy British' policy might well have had more success at a time of *rising* employment in the manufacturing sector, unpropitious economic circumstances are only part of the story. These same circumstances did not, after all, prevent the government in 1981 from discarding the tradition of relying on British designs of nuclear power reactors in favour of the United States' Westinghouse design of pressurised water reactor (PWR) for the Sizewell B power station, even if domestic employment was largely safeguarded through the type of licensing agreement negotiated (only 3.5 per cent of the manufactured content was scheduled to come from overseas). But curiously, what punctured the 'buy British' preference on the nuclear side of the electricity-generating industry helped to sustain the protection enjoyed by the defence-contracting industry. This was privatisation. The Conservative government's wish to sell the electricity-generating industry to private shareholders could only be carried out *in toto* if the nuclear side of the industry - about 20 per cent of the whole - could also be made to appear a commercially attractive proposition. A step towards that involved equipping it with the most cost-effective power reactors. In the end, the sale of the nuclear part of the industry was aborted, but that is another issue.

The planned privatisation of the defence-contracting industry, on the other hand, reinforced, however temporarily, the 'buy British' policy of the Ministry of Defence. Privatisation was a comprehensive, protracted and broadly successful undertaking to sell off to private shareholders defence contractors that had in the past been nationalised or partly nationalised, as well as defence contractors that had never been in the private sector. The former included Rolls Royce (privatised in 1988), and British Aerospace (privatised in two phases, 1981 and 1985); the latter included the Royal ordnance factories (sold as Royal Ordnance to BAe in 1987). No one would be prepared to buy defence-contracting companies or shares in them if there had been any suspicion that they were about to be exposed to foreign competition and thus have three-quarters of their market threatened. The single-mindedness with which a dowry was put together for this particular bride is illustrated by the story recounted above of ALARM, BAe, Royal Ordnance and the fixed-price contract. With the benefit of hindsight, it is apparent that any thought that the government might actually have awarded an important contract to an overseas competitor during the prolonged privatisation phase could therefore be consigned to the realm of science fiction. But this too, for the development-industrial circle, was a *deus ex machina*.

There are two other less transient and more structural reasons why the development-industrial level was in a good position to resist attack: global policy and 'industrial strategy'.

The global policy

In terms of global policy - principally the discreet security relationship between Britain and the Arab states of the Gulf, maintained since 1970 in spite of the formal British withdrawal from the region - the industrial level of defence policy and the operational level are found to be reasonably compatible. Even so, they are not entirely in line. The relationship or alliance between the operational and development-industrial circles within the global policy cannot easily be subsumed under the simple idea of a 'military industrial complex'; there is too much tension within it for that.

Levels in league

The alliance or alignment that exists certainly goes well beyond the 'floaters' of the previous Section, with the armed services doing for arms exporters what mannequins do for a couturier. The arms trade is a risky business but, as we have already seen, selling arms into regions where there is already a British military presence, even of a discreet kind, is one way of reducing the risks involved. Less loftily, any success the development-industrial circle may have in selling arms to semi-client states of the Gulf inevitably requires a special kind of after-sales service by the operational circle, in supplying training teams on site and training programmes for nationals of the purchasing state back in Britain. These training activities in Britain or overseas expose elite nationals of the recipient state to the ethos and assumptions of the British armed forces. This in turn increases the chances that future purchases of armaments may also be made in Britain and perhaps reduces the likelihood that the future political direction taken by the state in question will be out of keeping with British interests.

These two aspects of global policy are coordinated by two slightly shadowy organisations within the ambit of the Ministry of Defence: the Defence Export Services Organisation (DESO) and the International Military Services (IMS) organisation.[86] Both employ serving (DESO) or retired (IMS) military personnel, as well as civilians. Whilst there is apparently overlap (possibly not always constructive) between their functions, DESO seems to be primarily concerned that the development-industrial circle should be given every assistance (including that of the operational circle) in the promotion and sale of its products to overseas customers, whilst IMS has the job of making itself available as an agent for the convenience of potential customers for British arms (including the provision of financial assistance). DESO is about service to suppliers; IMS, service to potential customers. DESO may perhaps also represent something else besides. The head of that organisation seems invariably to be recruited on a temporary basis from the development-industrial circle (Sir Colin Chandler, whom we have already met, ran the DESO from 1985 to 1989 on secondment from BAe, who continued to pay his salary). It might not be too far-fetched to suggest that DESO, if not a Trojan horse, is a sort of enclave of development-industrial territory within the declaratory circle.

Divisions

And yet, as foreshadowed above, the global policy alliance between the operational and development-industrial levels is a rather odd one, both unequal and uneven. It is easy to see what the development-industrial circle gets out of the connection: cash flow, jobs and profits, even if it does not always get its own way. The complaint of J. McGowan, writing just before becoming Director of Marketing at Plessey Defence

[86] A good account of the DESO and IMS is given by J. McGowan, 'The Future for UK Defence Exports' in *Seaford House Papers 1986* (ed. Giles Binney), Royal College of Defence Studies, London, pp. 137, 154. There is a useful earlier account by Sir Ronald Ellis, the then head of the Defence Sales Organisation (which became the DESO in 1985; IMS becoming International Military Services as opposed to International Military Sales): 'Defence Sales', *RUSI Journal*, Vol. 124, No. 2, June 1979, pp. 3, 9.

Systems in 1987, is typical, blaming some of the difficulties facing British arms exporters on a 'reluctance by the Service Chiefs to compromise the design of platforms and weapon systems to allow them to meet a wider range of world roles'.[87]

On the other hand, it is hard to see how the operational circle, with the very important exception of the navy, benefits from the connection at all. The navy has the considerable compensation that its global policy role gives it a familiar and comfortable place in defence policy which the decline of Empire and the nuclearisation of great-power war planning were threatening to call into question. No such compensation is apparent for the army or air force. It is not as if the domestic defence industrial base has really played the part of 'bespoke tailor' to the armed services' needs for armaments, providing the forces with equipment of a kind and to a timetable that could not have been obtained elsewhere. The widely asserted claim to the contrary - for instance General Perkins' assertion that the defence industrial base enables the British armed services to acquire precisely what they need - is plainly nonsense. All three branches of the armed services have had to make sacrifices on behalf of the development-industrial circle. It may be noted that the other popular development-industrial circle argument cited by General Perkins - that Britain needs a comprehensive defence industrial base to prevent its being subject to embargoes by foreign suppliers - is also specious. If Britain can tolerate dependence on an ally (the United States) for the supply of key components for its strategic nuclear force, surely it can do likewise where conventional weapons are involved, particularly so when imports are likely to be available from more than one source country.[88] In some ways it is easier to see what sacrifices the armed services have had to make to sustain the defence industrial base, than those they would have to make were the base allowed selectively and judiciously to erode.

For example, we have seen how the RAF preferred the US-produced HARM missile to the ALARM being offered by British Aerospace but were forced to accept the latter, thus suffering both a cost penalty and a considerable delay in the missile's becoming operational. One of the

---

[87] J. McGowan, 'The Future for UK Defence Exports', *Seaford House Papers 1986* (ed. Giles Binney), Royal College of Defence Studies, London, p. 139.

[88] The author is indebted to the article by Miss C. S. Rycroft *op. cit.*, p. 24, for this argument.

reasons officially advanced for Ministers' overriding RAF misgivings and ordering ALARM was because the sale of Tornado aircraft to Saudi Arabia negotiated by British Aerospace included the ALARM missile as part of the package.[89] In the same connection, an academic commentator, Joanna Spear, draws attention to a press report that, as part of this same export deal, the RAF had to allow the diversion to Saudi Arabia of 20 Tornado aircraft in 1985 which had been intended for its own immediate use.[90] Even the cancelled Nimrod programme seems to have disadvantaged the RAF. According to another academic commentator, Peter Byrd,[91] the original decision taken by the Callaghan government in 1976 to develop the British-built Nimrod 'airborne early warning' aircraft was in opposition to the wishes of the RAF who wanted the Boeing AWACS instead. Of course they eventually got the AWACS, but fewer than the figure originally thought necessary and ten years late! We have also seen how the army was fobbed off with little better than 'badge-engineered' export surplus tanks after the collapse of the Shah's regime and the Iranian order for modified Chieftains. Even the navy has been affected. There seems to be no public record of the navy's attitude at that time, but the Ministerial decision to buy the Spearfish heavyweight torpedo offered by Marconi/GEC (in preference to a cheaper US model) was, as we have seen, in part influenced by the prospect of being able to export it. Superficially at least, the navy paid for this through receiving 20 per cent fewer torpedoes than it might otherwise have done for the same outlay; put another way, the navy could have had the planned number of torpedoes *and* a new frigate for the same cost.

This tension in the alliance between the operational and industrial circles may be part of the explanation for the well-known tendency of defence-contracting firms to recruit to their boardrooms and public relations departments recently retired senior members of the armed forces. The alliance is thereby reinforced at its potentially weakest point.

---

89 House of Commons Defence Committee, Fifth Report, Session 1987-88, p. 33.

90 Joanna Spear, 'Relations with the South', in Stuart Croft (ed.) *British Security Policy*, Harper Collins, 1991, London, p. 195.

91 Peter Byrd, 'Defence Policy', in Peter Byrd (ed.), *British Foreign Policy Under Thatcher*, Philip Allan/St Martin's Press, Oxford, 1988, p. 174.

Global policy dividends

Besides helping to keep a goodly and reasonably predictable share of the foreign exchange reserves of the Gulf rulers in London, global policy pays in two other ways. First, it aims to push up the level of British arms exports to the semi-client states of the Gulf beyond what otherwise might be the case. Up to a point - Britain does not possess a monopoly of outside influence on the Gulf Arab states, with the possible exception of Oman - global policy provides or, more accurately, aspires to provide, an extension abroad of the captive market for the products of the industrial circle that the Ministry of Defence provides at home. Secondly, non-military exports to customers for arms might be slipped in on the coat-tails of the arms salesmen. The net extra income to Britain from the profit on these two categories of exports can be thought of as helping to offset the cost of maintaining a global navy.

Profits from arms exports can be estimated, albeit crudely. We know that arms exports to all customers account for about 25 per cent of the output of the development-industrial circle. The average profit made on these exports is not publicly known: quite possibly there are no consolidated figures on record anywhere. Defence sales at home are claimed to yield a profit of no more than 10 per cent.[92] Sales abroad are reputed to be significantly more profitable. If we generously supposed average export profits to be 50 per cent, profits on the overseas sales of the industrial circle work out at 15 per cent of the procurement budget. Not all of this income should be credited to the global policy; some exporting, in particular from that part of the development-industrial circle whose products were already internationally competitive, would have occurred without it (Germany, without a global policy, is a successful arms exporter).

In fact this calculation is over-generous to the development-industrial circle in one other way. The growing tendency for importers of arms to ask for and to receive some compensation from the seller in the form of offset has spread beyond transactions involving developed countries. Saudi Arabia now requires some proportion of the cost of an arms deal with Britain to be spent in Saudi Arabia itself, which is not necessarily

---

92 Major General Ken Perkins, *op. cit.*, p. 37. Figures for BAe profits 1990/91 indicate only 10 percent for all defence sales, home and export (*Sunday Times*, 29 Sept. 1991, 3, pp. 8, 9).

always the most cost-effective, and hence profitable, way (to the supplier) of meeting the contract. The other tendency, to aim to pay in kind - for instance in oil shipments - rather than in hard currency, also takes some of the gloss off profits.

The other source of profit for global policy derives from the possibility that a successful arms export deal might prompt the customer to look to Britain as a source of non-military imports as well. Thus J. McGowan:

> Defence sales are almost invariably high profile events in the countries concerned. The political and trading links developed can help to promote British foreign policy objectives and economic advantage.[93]

It is evident that for there to be indirect 'economic advantages' of arms sales, the importing country, or the type of non-arms import under consideration, must normally be of the kind where importing decisions are taken by central government. This reduces the scope for coat-tailing on arms sales immediately. On the other hand, recalling that (as far as can be judged from published information) British exports of arms are overwhelmingly within the aerospace sector, a successful negotiation concerning military aircraft could plausibly trigger subsequent deals on civil aircraft, since purchasing decisions on passenger versions of the latter are taken centrally in a wide variety of countries. The affinity between the military and the civil need not be quite so obvious. A successful sale of arms involving the military sales division of British Aerospace or GEC could beget civil export orders involving motor cars from the BAe Rover factory or power-generating turbines from GEC's heavy electrical engineering division. It also has to be said that the tendency of British exports of arms to be confined to Middle East recipients rather increases the chances that importing countries will be of the kind where decisions about imports are taken centrally.

Absence of reliable data on the magnitude and direction of British arms exports unfortunately precludes statistical tests of association between the conclusion of an arms deal with a particular recipient and any subsequent increase in the British share of that recipient's imports generally. Consequently analysis has to rest upon less solid foundations. The Solomonic judgement of Sir Anthony Parsons, former British

---

93 J. McGowan, *op. cit.*, p. 139.

ambassador to Iran, provides a convenient starting point.

Writing for publication in 1988, Sir Anthony acknowledges that Britain has interests to protect in the Middle East which include 'her greatly increased export market'. After observing that Britain's share of exports to the region was nonetheless smaller than that of Japan, United States, Federal Republic of Germany, France and Italy (in that order), he adds:

> in so far as state activity is a factor in export promotion, the government has performed effectively. The fact that Britain has not gained ground against her competitors is more a reflection of the relative performance of the British economy than of any inadequacy in government policy.[94]

Global policy (Parsons's 'state activity' would surely be subsumed within global policy, but it specifically refers here to the export-promoting activities of the Foreign Office), then, has done something to protect the Middle East market share of British exporters but, according to Sir Anthony, this would scarcely have been necessary were British exporters internationally competitive.

Sir Anthony Parsons's well-known scepticism as to whether the resources of government, or at least the Foreign Office, should legitimately be put into the export promotion business does need to be kept in mind.[95] Moreover, Britain's market share in the Middle East as a whole is lower than its market share in the Gulf, where the focus of global policy lies. Even so, in 1988, Oman (the British 'protectorate') imported less from Britain than from Japan.

The balance sheet

A balance sheet can be drawn up only in the very roughest of terms.

Total British visible exports to the Middle East and North Africa in 1989 were worth about £6 billion. British invisible earnings from the same source (tourism, banking, insurance, etc.) were probably worth about the same. Subtracting the value of arms sales, about £2 billion, leaves earnings of £10 billion in total. If profit on this were about 10

---

[94] Sir Anthony Parsons, 'The Middle East', in Peter Byrd (ed.), *British Foreign Policy Under Thatcher*, Philip Allan/St Martin's Press, Oxford, 1988, pp. 82, 84.

[95] See, for instance, *Report from the Select Committee on Overseas Trade*, Vol. II, p. 158.

per cent, the element of national income involved was about £1 billion. Of course it would be ridiculous to credit the global policy with defending this entire sum from the depredations of the more economically competitive Japanese and Germans. But accepting Parsons's judgement that the British trading position in the Middle East owes something to 'state activity', if what was owed translated into a few hundred million pounds of profit not otherwise realisable, this represents another contribution to offset the cost of the global navy.

Be that as it may, two important caveats remain. Whatever global policy in the Gulf, and the discreet British military presence there in particular, might have done for British exports during the 1970s and 1980s, it was facilitating (free of charge) Japanese and FRG exports (and oil imports) by helping to maintain stability in the region and by allowing trading patterns to be broadly determined by the same market forces that were, according to Parsons, working to Britain's disadvantage.

Secondly, the global policy alliance between the navy and the development-industrial circle has delivered less than it might have promised. Global policy is failing if it cannot find markets in the Arabian peninsula for British-made main battle tanks and major combat warships. The failure to find such markets in turn weakens the political influence supposed to flow from sales of arms and hence global policy itself.

Industry, defence and research and development

The second structural reason the development-industrial level has remained out of line with the declaratory is slightly less easy to categorise, although it builds on the explanation favoured by Wilson, noted above, that the ambivalence of the Procurement Executive of the Ministry of Defence was the real problem.

Procurement Executive and R & D

The Procurement Executive is only part of the territory of the declaratory circle that shares a border with the development-industrial circle: its more anonymous neighbours, the Defence Research Establishments (REs), also share a border. In 1989/90 the MOD employed 24,200 civilian personnel engaged in military research and development, of which about 5000 were educated to degree level and might be described as defence scientists. According to ACOST, the function of the REs is to ensure that the PE remains informed as a customer about the products of the defence industrial base and to assist the PE in 'the ... maintenance of a proper balance [within the defence industrial base] of indigenous defence R & D and production capabilities'; ACOST goes on to say, however, that this is 'essentially an undeclared policy'.[96] There is nothing very 'undeclared' about maintaining the defence industrial base (although we have seen that defence statements have varied considerably in the apparent priority given to this aspect of policy) nor, in consequence, about maintaining indigenous defence R & D capabilities, without which there would be no defence industrial base. What is certainly undeclared is the PE, in concert with the REs, acting to maintain national R & D capabilities as a whole.

Military R & D in Britain represents a large proportion of all R & D carried out nationally. In 1989 it probably amounted to 20 per cent of the total (the derivation of this figure, together with a short survey of some of the difficulties and uncertainties connected with British research and development statistics, are contained in the appendix to this Section).

Wilson's claim that the development-industrial circle benefits from the ambivalence of the Procurement Executive, as an efficient purchasing agent for the armed forces on the one hand and as an agent of government industrial policy on the other, can be usefully investigated. Within the general ambivalence, there is a particular ambivalence in its joint R & D role with the REs. In addition to those functions outlined above by ACOST, the ostensible R & D role is to ensure that the weapons reaching the armed forces embody sufficient of the latest advances in science and technology for them to maintain their effectiveness on, and readiness for, the battlefield. But there is another role, again as an agent of government industrial policy. This is the specific one of sustaining the amount of R & D activity within the national economy as a whole.

The protection provided to the development-industrial circle by the ambivalent R & D role of the Procurement Executive in concert with the

96 ACOST, p. 63.

Research Establishments is, as we shall see, an instance of the part being greater than (or lending strength to) the whole.

Industrial strategy?

Where governments possess an industrial policy, they may also possess an 'industrial strategy'. The latter calls for government action to improve the working of domestic industry by intervening to correct or anticipate the operation of market forces, either for economic reasons, where the market is seen to be performing less efficiently than it should, or for political reasons, where outcomes created by efficient working of the market are politically unacceptable.

A domestic defence industrial base whose chief customer by far is the government gives that government some control over industrial employment and, less firmly, over the trade figures. But since the defence industrial base accounts for only about 10 per cent of industrial employment, and arms exports about 3 per cent of all exports, the importance of this control is very much at the margin. Admittedly, employment in the defence-contracting industries and the value of arms exports can come under a political magnifying glass when compared with the unemployment and trade balance figures respectively. Levels of unemployment and negative current account balance of payments figures are often considered to be electorally sensitive, so that governments may be reluctant to appear to be unnecessarily inflating either. But this would seem to be an even stronger reason why governments might find the process of removing protection from the defence-contracting industries transitionally painful - depending upon the stage of the economic cycle reached - than anything else.

The contribution of the PE in concert with the Research Establishments to an industrial strategy, however, is of a different order.

Why R & D?

As already noted, by some measures at least, the international competitiveness of British industry seems to correlate with the research intensity of the industrial sector in question. In addition, it would appear to be the case that international competitiveness at the national level - measured by a country's share of the world market for manufactures, say - is connected positively with the national level of expenditure on research and development. An academic specialist in the economics of technological change, Paul Stoneman, refers to there being:

> sufficient evidence ... that private and social rates of return to innovation are high and that trading success is at least partly determined by innovativeness.[97]

In a pure market economy, characterised by free flows of information and the absence of large, market-dominating companies, it is difficult to see how very much R & D activity would ever be embarked upon. Investigating novel products or novel ways of producing familiar products is both expensive and chancy. Market-dominated small companies, if not deterred by the chanciness of research and development spending, will be deterred by their inability to reserve for themselves a reasonable share of the proceeds of any successful innovation which they have paid for, but which others - their competitors - will also exploit. Patent protection is expensive and in itself no guarantee.[98]

Large companies can afford to hedge against the possible failure of an investment in the research and development of one new idea by promoting several R & D programmes simultaneously and using the successes to pay for the failures. Additionally, absent market conditions, large companies have the comfort of knowing that, whilst successful innovations might be copied by their competitors, the latter have only a small share of the market within which to exploit the innovation.

But size also discourages innovation. Companies that are large enough to be able to afford to pursue research and development are only going to do so if they are under competitive pressure, whereas market-dominating companies are normally (by definition) virtually immune to such competitive pressures.

More or less accidental imperfections in the market may encourage

---

97 Paul Stoneman, *The Economic Analysis of Technology Policy*, Clarendon Press, Oxford, 1987, p. 33.

98 Smaller companies, the very ones who need patent protection from competition, are reluctant, at least in Britain, to bother. A recent survey of small firms shows that only half patent their discoveries/inventions. *The Economist*, April 17th, 1993, p. 34.

companies to invest in R & D. Within the British pharmaceutical and chemical sectors, for instance, there are companies large enough both to finance innovative activity and to protect their innovations. They can afford the legal costs of acquiring and defending patent rights, while their geographical and linguistic distance from foreign competitors is itself a sort of safeguard against the pillaging of ideas (and/or scientific personnel) by competitors. At the same time they are not so large as to be immune from foreign competition in the market place at home or abroad, which in turn acts as a spur to innovation.

The defence budget as an aid to industry

Normally, however, it cannot be assumed that markets will fail in such a way as to encourage companies to spend on R & D, and governments usually feel constrained to intervene to make up the shortfall. The method of intervention practised in Britain, as in the United States, is through the defence budget, or rather that part of it spent on procurement of weapons.

The main defence-contracting industries in Britain are large by domestic standards. According to Bittleston:

> A handful of large contractors ... dominate the scene. Although 8000 companies are Ministry of Defence 'approved', the industry exhibits the classic pattern in which 20 per cent of suppliers win 80 per cent of the available market ... in 1986-87 the MOD paid only six contractors more than $425 m, whereas 44 companies earned between $8.5 m and $17 m.[99]

As we have seen, the main contractors face little or no domestic competition and at the same time are protected from foreign competition. Consequently, left to themselves, they will under-invest in research and development, leading to a deterioration in the effectiveness of the weapons and weapons-platforms they produce. This laxity therefore has to be minimised - the operational circle would tolerate no less. It is the function of the REs, through their own 'intramural' efforts and their support via the Procurement Executive to the development-industrial circle, to keep standards high. In so doing, the outcome of the R & D

---

[99] Bittleston, *op. cit.*, p. 37.

work they perform and help inspire is available to be appropriated by other sectors of British industry.

There are a variety of other routes government could choose to subvent investment in industrial R & D. Doubtless some might be more efficient than the defence route, where official secrecy at the operational and development-industrial levels is bound to obstruct the smooth running of the 'conveyor belts' which move assistance for research and development in kind and in cash from government to the defence-contracting industries and, less predictably, from the latter outwards to national industry as a whole. For governments wishing to claim they have no industrial strategy (though not entirely convinced that this is unnecessary), the ambiguity provided by the defence route has its advantages. But in this case ambiguity has another function. International trade, whether within the European Community or amongst advanced industrial countries generally, is bound by certain conventions governing fair and unfair trading practices. It is often very useful for governments to be able to claim that no special favours are being advanced to their domestic industries. However, using the defence budget to improve the competitiveness of domestic industry is, by custom, permitted.

The scale of R & D support to industry coming from the defence budget cannot be properly appreciated by looking at the consolidated figures alone. For instance, the ACOST study of British defence R & D, referred to above, found that about 15 per cent of defence R & D spending in the electronics sector possessed transferable potential, i.e. had applications beyond the defence sector. (For aerospace they found the figure to be 20 per cent.) In 1986/87, R & D spending by government on defence electronics possibly came to about £500 million or 20 per cent of the total. The transferable amount was correspondingly £75 million. The Alvey Programme of direct government subvention of R & D in information technology was set at £40 million a year for five years; this was presumably thought by its sponsors to be large enough to have a worthwhile impact.[100] Even allowing for the fact that information technology is not the whole of electronics, the defence-derived figure and the Alvey figure are

---

[100] Government expenditure on the Alvey programme between 1985 and 1990 was £210 m. See K. Guy, L. Georghiou et al, *Evaluation of the Alvey Programme for Advanced Information Technology*, HMSO, London, 1991.

comparable although, of course, the former is permanent while the latter was temporary.

Similarly, whilst common sense would seem to suggest that government support for industrial R & D would have more impact if targeted directly at those recipients who would make the most productive use of it - rather than relying upon more or less haphazard leakage or 'spin off' from defence-contracting industry - there would be a number of difficulties with such schemes. Even assuming a way could be found that did not impinge on international trading obligations, it might be difficult to replicate the breadth of specialisms of the defence sector, whilst retaining the advantages of concentration on a single task or closely related set of tasks.

Procurement Executive as sub-contractor

Wilson's diagnosis, that the development-industrial circle benefits from the ambivalence of the PE, can thus be accepted in a slightly recast form. It is not so much the role of the PE as a 'shadow' Department of Industry that matters: the resources at its disposal seem too small relative to the size of the economy as a whole. This is not to deny that being able to award defence contracts to depressed regions and even to time defence contracts so as to have a counter-cyclical effect on the economy remain options that governments might be unhappy to relinquish. Even so, the important thing is the fact that the PE (in concert with the REs) possess the sub-contract, so to speak, for the transmission of government support for innovation to industry. The development-industrial circle is protected, then, partly because it stands at the end of one conveyor belt, in receipt of government support for innovation, and at the start of one or a number of others (even if they are of only limited efficiency) transmitting new ideas and techniques to British industry as a whole. Were defence procurement to be opened up to international competition, part of the defence industrial base would disappear and, with it, parts of the conveyor belt.

British defence budgets are inflated, then, partly because of an illusion of presentation. Some of what appears in the defence budget is government support for industrial R & D that would very probably still need to be made available even if the defence budget were severely cut. It is possible to estimate the likely sums involved, even if only in a very sketchy way.

The appendix shows that in 1989 the defence-contracting industries performed R & D to the value of about £1600 million. Opening up defence procurement to international competition would probably leave the aerospace part of the defence industrial base more or less intact, whilst making considerable inroads into the somewhat smaller naval and land equipment sectors. The proportionate drop in government R & D expenditure (as naval and land equipment 'conveyor belts', in particular, shrank) would be of the order of several hundred millions. If, as seems certain, government continued to value the 'industrial strategy' benefits of promoting investment in research and development in British industry (and contributing to its cost), something like this sum would still have to be found from somewhere else. In spite of the inefficiencies of supporting general industrial R & D through the defence budget, it might be rash to assume that whatever new method the government employed would be more efficient (apparent gains in efficiency might be compromised by a requirement to conform to the government's international trading obligations), so that the sum concerned could remain at several hundreds of millions sterling (1990 prices) annually.

Summary

There are a number of interwoven contradictions concerning the development-industrial level of British defence policy but, as befits a topic to do with money, they all boil down to a question of cost. These can be encapsulated as follows: the stated policy of 'value for money' in the procurement of defence equipment; the concentration (at least 90 per cent) of conventional procurement orders from the Ministry of Defence on the British defence-contracting industry; and the strong evidence that better value for money could be obtained by placing orders for warships and possibly tanks too with suitable foreign manufacturers rather than with home producers.

The reason for the contradictions persisting has something to do, of course, with the lobbying power of the development-industrial circle, and with the reluctance of government to test the truth of Churchill's dictum, that giving 'government patronage to foreigners' will simply not be understood by the electorate. It has something to do, too, with the plain fact that international trading regimes do not harry governments to expose their defence-contracting industries to international competition in the same way that they may be pressured over other national industrial activities. It also has something to do with the global policy. Influence over foreign governments enables arms to be sold to them; being able to sell arms extends influence. It would be difficult to know how to divide up the credit for the maintenance of British influence on the Arab shore of the Persian Gulf as between the armed services and the arms exporters. In addition, whilst the defence-contracting industry in Britain may sometimes sound as if it believed that manufacturing was still something that British industry did outstandingly well, and that during the Cold War years the country should have been preparing for a new major war much along the lines of World War Two, 'anachronistic' would not be an entirely fair description of its stance. Like the defence-contracting industries of the United States, but of course on a much smaller scale, it is closely involved in a tacit contract with government whereby public funds are channelled through them and made available for the national support of industrial innovation.

Wages of failure

The price of failing to bring the declaratory and operational levels into line, as we saw in the previous Section, included the war in the Falklands. The price of protecting the development-industrial circle from virtually all foreign competition has been a good deal less dramatic: it has simply cost money.

How much precisely is very hard to say. The cost of protecting the British defence-contracting industries is possibly somewhere near the Hartley figure of 25 per cent of the procurement budget. In other words, a policy of obtaining armaments from the cheapest politically acceptable source, which would by no means exclude all British manufacturers, would save about 25 per cent on the procurement budget. That startling savings can be made by 'judicious overseas purchase' or 'buying off the end of a long production run' is something British governments have been made well aware of in their years of purchasing nuclear delivery systems and related equipment from the United States.

Not all of these savings would have been realisable during the period of our study. Indeed, it might have been sensible to keep some sort of strategic hedge against the possibility that a prolonged war with the Warsaw Pact countries could have been fought without the nuclear threshold ever being crossed. Thus, instead of buying off the end of someone else's long production run, one sort of strategic insurance in relation to selected armaments which British manufacturers cannot build efficiently would be to negotiate construction (like the Sizewell B nuclear power station) under licence from a foreign manufacturer. The cost of this would normally be in between that of direct overseas purchase and wholly domestic production.

As we have seen, quite a large proportion of the British defence-contracting industries would in fact survive a purge of this kind, particularly the aerospace sector. Thus the contribution of the development-industrial level to the tacit global policy would in turn remain largely unaffected. Global policy is not furthered by arms that cannot be sold, and the record of British sales abroad of major items of land and naval equipment, particularly in the second decade of the span of this study, is so poor that any restrictions on re-export (placed by whichever non-British source supplied major weapons to the army and the navy under the new regime) could hardly be described as irksome.

On the other hand, the tacit contract held by the defence-contracting industries as middle men in the conveying of government aid to industrial R & D is a more substantial guarantor of the status quo. A redefined defence industrial base concentrated more on and around aerospace would halve the capacity of the R & D conveyor belt.

Appendix: UK R & D 1989

The matrix below (Table 3.1) is derived from a number of sources. The most important is the similar matrix, but without either government departments or industry broken down into civil and defence sectors, published by the Cabinet Office (*Annual Review of Government Funded Research and Development 1991*, HMSO, London, 1991). The same source gives valuable estimates for the fraction of R & D undertaken by the defence-contracting industries which they have financed themselves, and the fraction paid for by government and others. Similar estimates are given for the civil industries.

Very useful additional guidance is given by the National Audit Office publication *Classification of Defence Research and Development Expenditure* (Report by the Comptroller and Auditor General, HCP 105, HMSO, London, 1991). The NAO highlights the chief problem connected with constructing matrices of this kind: inconsistencies between government figures for what it claims to have spent on R & D undertaken by industry and the figures (normally smaller) industry itself gives for the value of the R & D which it has performed for government. Thus our matrix - which, following the NAO and the Cabinet Office, takes industry's figures for the value of R & D work carried out - shows the defence-contracting industry as having performed R & D for the Ministry of Defence to the value of £940 million in 1989. The MOD claims, however, that it spent £1152 million (Cabinet Office figure).

The reason for the discrepancy is not entirely understood. Confusion over the inclusion or otherwise of Value Added Tax is one possible explanation. Wrongly classifying some defence R & D in the 'grey area' between development and production may be another. This problem is not unique to Britain.

It is because of these uncertainties that all figures here are cited to the nearest £5 million though some errors will nonetheless be greater than this allows for. The figures marked * are particularly rough-and-ready estimates although, being small, they do not affect totals very significantly one way or the other. The agreement between the above matrix and the more consolidated Cabinet Office version is quite close (the Cabinet Office figure for the total value of R & D performed in Britain in 1989 is £11531.8 million, as against our £11525 million).

The total amount of defence R & D performed in Britain in 1989 can be defined as the sum of that paid for by the Ministry of Defence and that paid for by the defence-contracting industries. This comes to about £2390 million, about 20 per cent of the total amount of R & D performed in the country.

Table 3.1

UK R & D 1989

	Performer £m							
Funder £m	Gciv	Gdef	HE	Civ Ind	Def Ind	Other	Total	O'seas
Gciv	425	50	1315	300	-	205	2295	120
Gdef	20	940	20	-	940	10	1930	120
HE	-	-	80	-	-	-	80	-
Civ Ind	115*	20*	110	4875	-	180	5300	-
Def Ind	15*	5*	20	-	400	20	460	-
Oth/O's eas	35	50	245	775	280	75	1460	-
	610	1065	1790	5950	1620	490	11525	Total

Key

Gciv: government departments, excluding defence
Gdef: Ministry of Defence
HE: higher education establishments
Civ Ind: the UK civil industrial sector
Def Ind: the UK defence-contracting industries
Oth: performers not elsewhere specified
O'seas: overseas performers
Oth/O'seas: overseas funders and funders not elsewhere specified
*: see text

Conclusion

Introduction

This book set out to answer three questions about British defence policy in the two decades since 1970.

One concerned the apparently inflated size of the British defence budget when set alongside the defence budgets of comparable European members of the NATO alliance. The other was essentially to do with the Royal Navy and simply asked what it was for, since British defence policy in NATO did not seem to envisage a great-power war of sufficient duration for the British navy to have very much bearing on the outcome. Neither of these questions is particularly original. The new Labour government publicly asked the first of them in 1975. The new Conservative government publicly asked the second in 1981 (echoing Duncan Sandys's earlier enquiry along the same lines a quarter of a century earlier). The third question was about the British defence-contracting industry and its mysterious survival during a period of general run-down of British manufacturing capacity.

Two linked approaches were taken to arrive at an answer to the defence budget question. The implicit, indirect approach - which in fact tackles both questions together and forms the major part of the book - and the explicit, direct approach, employing statistical analysis, which for that reason is presented in the form of an annex.

The direct approach

The direct approach first establishes that there is a case to be answered. The degree to which the British defence budget, expressed as a proportion of gross domestic product, has exceeded the NATO European mean in the period 1970 to 1990 averages at about 40 per cent - in round figures, about 5 per cent of GDP for Britain as against a European NATO average of about 3.5 per cent. In other words, if Britain spent the same proportion of its GDP on defence as the NATO European average, its defence budget would typically have been about 30 per cent smaller than it actually was. More precise calculation shows the figure to be 28 per cent. In 1990, this would have translated into a difference in actual expenditure of £5800 million - still too large to be explained by obvious special circumstances. In fact, neither Britain's status as a nuclear power nor its military involvement in Northern Ireland significantly reduces the excess. However, once account is taken of the fact that Britain uses a system of voluntary recruitment to its armed forces as opposed to the conscript method favoured by nearly all its European allies, the apparent excess reduces a little, from 28 to about 25 per cent.

The annex tests the suggestion that there were two fundamental reasons for high British defence budgets. One was Britain's peculiar position as an original major contributor to the NATO alliance and (like a smaller version of the other major contributor, the United States) becoming over time a hapless victim of the inbuilt tendencies in such alliances towards uneven 'burden-sharing'. This occurs because smaller members of multi-member alliances can reduce their contributions to the common defence with less fear that, by so doing, they might jeopardise its whole structure than is the case with larger contributors trapped into a more *communautaire* posture. This structural factor accounts for about 10 per cent of the British defence budget, or two-fifths of the apparent excess.

That alliances are prone to this sort of structural distortion of contributions to the collective defence has been known for a long time, although the focus of previous investigations has usually been on the more marked plight of the largest contributor to NATO, the United States. The second reason put forward in the annex to explain the relative size of the British defence budget emerges from the main text of the book. Here the suggested reason is that Britain continued to pursue, well after 1970 and the supposed withdrawal from East of Suez, a still expensive but now tacit and unofficial 'global policy'. Global policy is used in this book as shorthand for the employment of military and related methods for the promotion of specifically British interests of a broadly mercantile kind, mainly on the Arab shore of the Persian Gulf. The statistical enquiry puts the cost of this tacit global policy at about 15 per cent of the defence budget. This is larger, by three times, than the only figure ever advanced by government for the portion of the British defence budget not officially allocated to NATO tasks.

The indirect approach

The indirect approach to answering questions about the size of the British defence budget and the function of the Royal Navy is in some ways at least as elaborate as the statistical approach, but much more finely grained. It is based on the idea, again not original, that there are three levels of defence policy, each with its appropriate circle of members: the stated or declaratory, chiefly the domain of politicians; the operational, which is occupied chiefly by the armed services; and the development-industrial, which is mainly populated by the defence-contracting industries.

The navy is as it is simply because that is what the navy wants - a global-policy navy or blue-water navy. Its role within stated or declaratory policy is very hard to explain because, in size and in composition, it is very different from the one that would exist were the declaratory and operational levels of policy to be perfectly in line, which they are not and have not been since at least 1970. John Nott (later Sir John) attempted in 1981 to create the sort of NATO-orientated navy that would have matched stated policy better, but failed. In doing so he gave a rough indication of how much cheaper the new navy would have been than the old. In very round terms, something like one-tenth of the defence budget was involved.

Bearing in mind that Nott did not envisage Britain surrendering its vestigial, declared out-of-area activities, conventionally costed at 5 per cent of the defence budget, the total sum really involved - the cost of the global-policy navy plus 5 per cent- is exactly the 15 per cent figure we are looking for. In reality Nott did envisage some small extra expenditure on the other two branches of the armed services to offset the planned loss of some of the navy's out-of-area capability, so that the exact agreement between the two figures is, for this and other reasons, rather fortuitous.

It is not only the operational level that was out of line with the declaratory during this period; so was the industrial level. One important instance can be made to stand duty for others. Stated British policy with respect to the purchase of military equipment has long been getting value for money. At the same time, the industrial circle has succeeded in protecting itself from international competition to the extent that less than 10 per cent by value of Ministry of Defence purchases of conventional military equipment is imported (this compares with the 30 per cent of all manufactures within the British economy met by imports). The 10 per cent figure has remained essentially unchanged over the entire twenty-year period of our investigation. For the industrial circle to have achieved this feat 'legitimately', it would need to have been among the most efficient manufacturers of armaments in what used to be called the Western world. The evidence indicates the contrary. Aerospace manufacturers alone seem to be efficient by international standards. Producers of land equipment (tanks, artillery) appear to be less efficient than at least some international competitors. But British builders of warships take the wooden spoon. Whereas British manufacturers of main battle tanks have managed some export successes between 1970 and 1990 (although, perhaps significantly, none in the second decade), no major new British warship has ever been exported in the entire twenty-year period.

Thus another contribution to the size of the British defence budget comes from the 'buy British' premium that the industrial circle has managed to persuade the ministerial segment of the declaratory circle to pay. Estimating the size of this premium is particularly difficult, and no official figure (made with privileged access to information normally screened behind the veils of commercial and military secrecy) seems ever to have been produced, but again we are probably dealing with about one-tenth of the defence budget.

However, we would not really expect this particular figure to show up in any realistic form in our statistical comparison of the defence budgets of European members of NATO. This is simply because the British defence-contracting industries are not alone in NATO in obtaining government protection from imports. A number of other European NATO members, France especially but not solely, are just as hostile to exposing domestic defence-contractors to international competition as Britain is (although this does not preclude the possibility that their defence industries may be more efficient than, or inefficient in different ways from, Britain's).

On the other hand, France is not entirely typical (the nationalistic and self-sufficient orientation of the development-industrial level of French defence policy, we may note in passing, is not so far removed from its stated policy of ambiguity towards NATO, as is the case with Britain). Smaller members of the alliance often have had little option but to import major items of military equipment. And even Germany, at any rate in the earlier part of the past two decades, was quite ready to look overseas for arms.

In other words, the self-sufficient orientation of certain other NATO members' defence-industrial policies usually allows us to ignore, to a first approximation, the possibility of the 'buy British' premium showing up in a formal comparison of NATO statistics such as appears in the annex. At the same time, it is a reminder that the statistical enquiry does not aspire to complete accuracy. Were it to do so, the fact that some NATO member states have quite regularly bought weapons from the cheapest available source, where Britain has not, would show up in the statistical analysis as a small factor (along with the larger, out of area, explanation) contributing to the 15 per cent by which the British defence budget typically exceeds the NATO European norm (once 'burden sharing' effects have been allowed for).

This book is not concerned with demonstrating that the British defence budget *should* have been smaller or larger by some particular amount than in fact has been the case. The extra costs incurred by the tacit global policy and the global-policy navy, it could be argued, were as nothing when set against its success in encouraging traditional rulers in the Gulf to keep a respectable share of their surplus oil earnings in the City of London. An argument of this kind could also have helped subliminally to secure the navy's position against Nott-style attacks. On the other hand it is not obvious that a different implicit diversion of British military activity from the NATO role to out of area could not have been devised so as to be more effective or cheaper than the global-policy navy was. Equally, the two-way relationship between global policy and the British defence-contracting industry seems to have been far from optimum. Arms exports do something to promote global policy, and the undifferentiated total of British arms exports into the Arabian peninsula appears to be reasonably high. But if global policy is also meant to do something for arms exports - by creating markets for British-made arms where none would otherwise exist - the record is much poorer. Virtually all British arms exports are in the aerospace category. Arms that cannot find overseas customers can do nothing for global policy; on the other hand if there are no foreign buyers, then global policy is equally doing nothing for arms exports. The fifty per cent of the British defence industrial base that is not aerospace only pays its way if it offers domestic value for money. This it does not do. Ministry of Defence purchases of British-made warships and land equipment, when allies offer better value for money in both categories, is what accounts for the 'buy British' premium which adds roughly one-tenth to the British defence budget.

The defence review

The process of bringing the different levels of defence policy into line, or, in other words, removing mutual inconsistencies, is another name for the defence review. The consequences of failures in alignment can be extremely serious. This is not simply a matter of possible waste of public funds. Because declaratory policy for the decade prior to 1982 was that Britain did not possess the means of projecting military power much beyond the geographical confines of NATO, Argentina was not deterred from invading the Falkland Islands. Too late, the Argentine authorities discovered their mistake: operationally Britain did possess the necessary forces, but the Argentines could hardly be blamed for not appreciating this before they embarked on their venture.

That a review is needed or may be long overdue is not enough: some significant external or exogenous event seems to be required to trigger it. This could be a financial crisis, a change in government or a reversal in the field for British arms - or any combination of these. Triggers can act in reverse as well. The British victory over Argentina actually rendered null and void a defence review that had been put under way by the new Conservative government.

In any case, executing a defence review successfully seems to be enormously difficult. The operational and development-industrial circles have the important advantage of inertia: if nothing happens, they have won. The declaratory circle, or rather those of its members of suitable political seniority, has to take the offensive and the initiative, the other two circles merely need to defend the status quo. In this way both the navy (if not the rest of the operational circle) and the development-industrial circle seem to have ridden out the nuclear revolution almost as though it had never happened. The navy resisted the Healey reviews of the late 1960s by a variety of stratagems, and it is by no means clear that Nott would have had his way either, even without the Falklands war. As for the development-industrial circle - the defence-contracting industries - that they managed to sustain their protected position during the Thatcher years - when exposure to competitive forces seemed almost to be the government's chief guiding principle - speaks of impressive powers of resistance.

Even so, whilst there is evidence of some degree of mutual support as between the operational and development-industrial circles acting to make the task of the declaratory circle even harder - crudely, both may be described as having an interest in seeing defence budgets kept high - this is a long way from being the seamless compact (a military-industrial complex) that has sometimes been alleged. For one thing, all three branches of the armed services have suffered from the inefficiencies of the development-industrial circle, especially when equipment has been expensive and slow to come into service compared with what could have been obtained from a foreign supplier. For their part, the defence-contractors would claim that the selfish concerns of the British armed services often force them to produce armaments that are too sophisticated to be easily marketed abroad. For another, within the operational circle, the navy can be singled out has having been pressed into making fewer adjustments to its institutional preferences than the other service branches where ground has had to be given. All of this, one might think, leaves obvious openings which could be exploited by shrewd managers of a future defence review.

The future: options for change

The 'three levels' framework of analysis has here been used to look backwards in an attempt to make sense of British conventional defence policy in the period since 1970. Has it anything useful to say about the future?

One testable, if not necessarily very useful, prediction can be made. It seems clear that in matters of defence policy, clean breaks with the past are almost impossible to effect. As we have seen, previous attempts to swing British defence policy away from an earlier to a new set of priorities have not always been very successful: that is to say, operational and industrial levels of policy have not kept in step with the declaratory. There is an almost mechanical reason for this - the large momentum imparted to old policy directions by the long lifetimes of modern weapons and the long gestation periods of new generations of these weapons. There is no sign whatever that either the lifetimes or gestation periods of weapons are getting any shorter. But more than technical factors are present. Institutional factors also matter. The navy battened down hatches and used every means to cling to an earlier role because it feared an actual or relative loss of status as a purely NATO-orientated navy compared to what it had formerly enjoyed. In turn, the defence contractors had jobs and profits to defend; to do this, they promoted the obsolescent idea of the strategic importance of national self-sufficiency, which interlinked neatly with their emergent modern role as conduits for directing government support to industrial innovation. Apparently paradoxically, multinational collaborative projects for the development and production of weapons, which may appear to compromise self-sufficiency (in fact, as we have seen, they do not) in fact add to the 'mechanical' momentum of earlier decisions by creating new political obstacles to subsequent reconsiderations of policy in the context of defence reviews or otherwise.

In essence, therefore, past levels of British defence expenditure are a good predictor of future levels, while increases in government spending on defence are easier to effect than decreases. It can be predicted quite safely that over the next five years Britain will continue to show a higher rate of expenditure on defence as a proportion of national income than the average among what may still be called NATO's European members.

NATO's uncertain future is an indication that these are particularly interesting times for British defence policy. For the first time in more than half a century, the British security environment does not contain a large military threat - a clearly identified potential foe of known and impressive capability. Instead of 'the threat', there are now simply a multiplicity of minor threats. Such a large shift in the external constraints on declaratory policy makes an eventual reformulation of that policy inevitable.

The curious thing about the defence review instituted by the then defence minister Tom King in June 1990, 'Options for Change', was that it avoided reformulating stated policy, whilst announcing proposed cuts in the size of the armed services. The absence of a new stated policy may testify to the truth of what has already been said in Section One about the difficulties that may attach to reformulating declaratory policy, in spite of its being 'only words': but it is likely to have real consequences.

It is an old saying that it is difficult to argue against something - the well dug-in ranks of the operational and development-industrial circles - with nothing. This may be slightly unfair in that, two years on from 'Options for Change', the 1992 Statement on the Defence Estimates makes a hesitant stab at a declaratory policy. This speaks of 'three overlapping roles' for British defence policy, but in such vague, general and unexceptionable terms as to suggest a stop-gap as opposed to an end product.[1] It is certainly no springboard from which the declaratory circle could reply to claims by the operational, say, that the ending of the Cold War has seen an increase in what is expected of the British armed forces (an increase in 'commitments' would be their phrase, e.g. the Gulf war and its aftermath) rather than a decrease. Even when there is a stated policy of reasonable intellectual coherence, making ground against the operational and development-industrial circles is, as we now know, hard enough. Without one, it is difficult to see how any progress can be made. This curious vacuum in 'Options for Change' and its poorly documented character (a statement to the House of Commons followed by glosses in subsequent statements on the defence estimates), plus the absence of evidence of any real struggle (such as the resignation of a junior defence minister 'gone native' within the operational or industrial circles), have led many commentators very reasonably to doubt its claim to be considered a defence review at all. From the

---

[1] The three overlapping roles are:

1. To ensure the protection and security of the United Kingdom and our dependent territories, even when there is no major external threat.

2. To insure against any major external threat to the United Kingdom and our allies.

3. To contribute to promoting the United Kingdom's wider security interests through the maintenance of international peace and stability.

Statement on the Defence Estimates, 1992, Cm1981, HMSO, London, p. 9.

perspective developed here, the failure to redefine stated policy simply means that difficulties have been avoided rather than faced. Leaving the field to the other two circles makes it almost certain that the huge Nott-sized cuts announced by King - of 25 per cent in the size of the army - will never be implemented.

Alternatively

It also has to be admitted that the method developed here may not prove to be a reliable crystal ball with which to read the future at all. King's 'Options for Change' may not be just the most recent (if rather unsatisfactory) example of a genre associated with the names of Sandys, Healey and Nott. Instead, it may be the first of a new kind of defence review, appropriate to the new strategic circumstances Britain finds itself in. The important fact about these new circumstances may be the much reduced risk of nuclear war. When nuclear war loomed large, the peacetime role of politicians in the making of defence policy - all defence policy - expanded: as nuclear war becomes more remote, so might the three-levels approach used here itself become less relevant. After all, it was borrowed from a particular approach used in studying the military nuclear policy of the United States.

Conclusions to books are said to be about picking up the apples that have been shaken from the tree.[2] It may be that the device borrowed here from Ball and Snow to allow a better purchase on the more awkward branches, having done its job, will be of no further use to anyone. Even so, the author allows himself to hope that the consumer may have found the fruit wholesome and the exercise to have been worthwhile.

---

[2] See J.W.N. Watkins, *Hobbes's System of Ideas*, Hutchinson University Library, London, 1964, p. 163.

Annex: The Size of the British Defence Budget

Introduction

If the three levels of defence policy as described in the main part of this book are out of line, this should show up in the figures relating to the size of the British defence budget. Declaratory strategy will not by itself be sufficient to explain the size of this budget. More particularly, if the budget were appreciably larger than the stated policy would seem to require, this would be evidence that the defence-contracting industries had provided the operational circle with a military capacity in excess of that set out in declaratory policy. Almost unwaveringly declaratory policy has over the past twenty years described Britain's armed forces as nearly totally dedicated to NATO, from the 1973 statement that 'virtually the whole of the RN, most of the combat units of the Army and the majority of the combat a/c of the RAF are committed to NATO',[1] to that of 1976 that 'almost the whole of Britain's defence effort is now devoted to supporting NATO's strategy'[2], through to 1979 when it was stated that 'the UK commits the great bulk of its forces to

---

1 *Statement on the Defence Estimates*, Cmnd 5231, HMSO, London, 1973.

2 *Statement on the Defence Estimates*, Cmnd 6432, HMSO, London, March 1976.

the Alliance'[3], to the gratifyingly quantitative 1987 (and 1988) claim that 95 per cent of Britain's military capacity is devoted to NATO.[4] Thus an estimate of what the British contribution to NATO should cost can be compared, at least in principle, with what actual British defence expenditure is.

There are a number of difficulties involved in making an estimate of this kind in practice. A reasonable initial step would be to compare British defence expenditure with that of comparable members of NATO; that is to say, fellow member-states whose historical concern at the threat to their security from the East was as least as great as that of Britain and whose size and degree of economic development might be expected to have about the same shaping effect on their commitment of resources to defence as would be the case with Britain. In other words it is natural to compare Britain's level of expenditure on defence with that of France and Germany or even, with the growing strength of the Italian economy in mind, Italy too. The incoming Labour government seems to have been alluding to comparisons of this kind when it spoke in 1975 of its plan to reduce the British defence budget to something like the average for European members of the alliance, its Tory predecessor having put into reverse, it claimed, the Labour decision in 1968 to focus Britain's military activities on Europe.[5] A consequence of the line of reasoning developed in Section One is that, had this plan been implemented (which it was not), it would have required a real defence review and a total realignment of the three levels of policy.

Data doubts

Even if the reasonableness of international comparisons of this sort is accepted, other difficulties have to be overcome, not least those relating to data on defence expenditure themselves. The first of these concerns the trustworthiness of the data as published. Defence expenditure is directed at the operational and development-industrial circles, the two least open and public of the three identified in the main text; thus it might be suspected that the published level of defence expenditure in any

---

3 *Statement on the Defence Estimates*, Cmnd 7474, HMSO, London, February 1979.

4 *Statement on the Defence Estimates*, Cm 101-I, HMSO, London, 1987 and *Statement on the Defence Estimates*, Cm 344-I, HMSO, London.

5 *Statement on the Defence Estimates*, Cmnd 5976, HMSO, London, 1975.

one year may not be the true figure. In the case of Britain doubt on this score is put aside as far as the total global figure for the defence budget is concerned, especially in terms of historical figures and not current figures (small retrospective corrections are not unusual). The actual apportioning of spending under the separate headings of the defence budget, on the other hand, even in peacetime cannot always be assumed to represent an honest attempt to convey the truth. The cost of the Chevaline programme to provide a mid-life update to the re-entry vehicles of the Polaris force, chiefly to provide greater assurance that Britain's comparatively small number of strategic nuclear warheads would be able to penetrate Soviet anti-missile defences, was hidden for a time under the budgetary heading of 'maintenance'[6].

Similar doubts about the trustworthiness of published data can be legitimately expressed with regard to the defence budgets of other NATO countries too. Indeed a recent painstaking NATO-sponsored survey of defence expenditure within the alliance since 1970 rather disquietingly describes its data on French military expenditure as 'indicative only'. However, this does not prevent the authors of the survey from quoting French figures to the same apparent level of precision accorded those of other member states. Additional difficulties present themselves when an attempt is made at international comparisons. Whilst it is possible to adopt a standardised version of what constitutes defence expenditure, more than one standard is feasible. Moreover, because there is legitimate variation in how national expenditure over fiscal-year periods should be converted to a common calendar basis and similar variation in how to convert to a common currency - the latter being necessary, for instance, when an attempt is made to calculate a European NATO average defence expenditure figure - it is not surprising that precise equivalence is very rare between any two authoritative estimates of how defence spending varies among NATO states.

However the situation is not beyond salvage. The fact that authoritative estimates of NATO defence expenditures vary can be useful in judging how much confidence to attach to a single set of published data for NATO countries. The method relies on the fact that four separate authoritative calculations of the defence expenditures of NATO countries are publicly available. Since these assessments are based on

---

6 Michael Dockrill, *British Defence since 1945*, Basil Blackwell, Oxford, 1988, p. 107.

a reasonably uniform methodology which aims to render individual nationally published figures comparable on a NATO-wide basis, a comparison of the figures derived by each authority will indicate what sort of margin of error (or what confidence limits) should be attached to any given listing.

The shared methodology depends on a so-called NATO 'standard definition' of defence expenditure which in practice, for all NATO member states but two, produces a higher - sometimes substantially higher - figure for defence expenditure than the corresponding nationally published figure. The exceptions are Luxembourg, whose nationally published figure follows the NATO standard exactly, and Holland, whose national definition of defence expenditure is even more widely drawn than that of the NATO standard.[7]

The four authorities employed in the present survey are the NATO secretariat (NATO), the United States Department of Defense (Pentagon), the International Institute for Strategic Studies (IISS) and the Stockholm International Peace Research Institute (SIPRI). NATO, naturally, employs the NATO standard definition of defence expenditure *tout court*; the IISS supplies figures calculated on the same basis (as well as its own estimates). The Pentagon and SIPRI both base their method of calculation on the NATO standard, and whilst it is impossible to say how closely, the indications are that departures are slight.

Inspection of Table 4.1 (p. 185) enables a number of conclusions to be drawn. The first is that, with the exception of the south European member states (the Southern tier of NATO), agreement between the authorities is good. Even within the Southern tier, agreement may be better than shown because the SIPRI figure for Greece looks suspiciously divergent and may contain an error. Secondly, although the absence of sufficient data to carry out the NATO European average calculation for the SIPRI column is unfortunate (SIPRI gives no estimate for national GDPs in dollar terms), there is no difference between the other authorities as to the weighted European NATO average figure, which helps confirm the claim made above that each is employing essentially the same definition of defence expenditure.

Thirdly, a rule of thumb for the confidence interval or margin of error likely to attach to any one would-be comparable listing of data is plus

---

7 For a brief description of what the NATO standard means, see *The Military Balance 1990-1991*, Brassey's for the IISS, London, 1990, p. 9.

or minus one in the first figure after the decimal point, except for Turkey and Portugal and possibly Greece, where plus or minus three in the same figure may be indicated.

The above allows us to proceed to attempt to draw conclusions about how far British levels of expenditure on defence may have been in excess of the NATO European average in the period since 1970, whilst reasonably well equipping us to discriminate between a real difference and a statistical distortion. Table 4.2 (p. 186) shows that Britain has consistently exceeded the European average by an amount well in excess of the margin of error. This table, based on data published by NATO (hence the 4.1 per cent of GDP attributed to Britain in 1989), shows for instance that Portugal for the period 1980/89 and Norway since 1985 have consistently allocated resources to defence at about the NATO European average rate.

But there is still room for doubt. It could be argued that like is not being compared with like. Whilst the effects of different national accounting methods and different national conventions in the presentation of statistics relating to military expenditures have been overcome or at least allowed for, some important national differences may still remain. If anything like a fair comparison is being sought between British defence expenditure figures and those of its principal European allies, the most significant difference to be taken into account is manpower policy. Britain has an all-volunteer system of military recruitment, in place since 1960, whereas its continental allies (Luxembourg excepted) all employ some method of conscription.

In 1974 the British Ministry of Defence calculated that, were Germany to move to an all-volunteer system, its defence share of gross national product (GNP) would rise by about 0.3 percentage points, from about 3.5 per cent GNP to 3.8 per cent.[8] For France the upwards adjustment would be greater, at 0.6 percentage points (from about 3.8 per cent to 4.4 per cent), because of France's policy of paying conscripts particularly poorly.[9] However, the Ministry's methodology has been

[8] Gross domestic product (GDP) is the monetary value of all goods and services produced within a state. The gross national product (GNP) is the total income of all citizens irrespective of origin. GNP = GDP + net income from abroad (net income from abroad = profits, dividends, plus interest earned abroad minus income payable to foreigners). In fact the difference between these two measures of national wealth is often very small. For Britain the GNP/GDP ratio = 1.01 (*The Economist*, 21 Sept. 1991).

[9] *Second Report from the Expenditure Committee*, Session 1974-75, HC Paper 259.

described as vulnerable to criticism,[10] and a new survey of the question is offered below.

Adjusting for conscription

There is probably no 'best way' to create strict comparability between the defence budgets of NATO member states that rely on conscripts and those that do not. The method chosen here is deliberately conservative. It is assumed that armed service personnel take 12 months to complete their training - this figure seems to correspond to the modern experience of the British army.[11] If the conscription period is also 12 months long, then each conscript in service is on average half trained and therefore roughly half as useful as a fully trained person. Consequently, in these circumstances, one volunteer (when trained), whose term of service will usually be very much longer than one year (the training period), can take the place of two conscripts.

If on the other hand the conscription period is longer than 12 months, those conscripts beyond the twelve-month point are counted as fully trained and only replaceable with volunteers on a one-to-one basis. Calculating the cost of abolition of conscription, whilst essentially maintaining the fighting strength of the forces at its pre-abolition level, then entails knowing the salary (career average) necessary to attract the required number of volunteers. This will be about - again on British experience - twice the gross domestic product per head, and this figure, multiplied by the estimated number of volunteers necessary to make up for the disappearance of conscripts will normally lead to a new, higher defence budget (column i, Table 4.3, p. 187). The exception to this rule will occur when conscript costs are high, e.g., where they are paid a salary comparable to that obtainable in civilian life. Table 4.3 makes the simplifying assumption, not that conscripts cost nothing, but that they cost no more than the figure government deems as an adequate minimum unemployment benefit in the country concerned. Again

10 Paul Cockle, 'Observations on the proposal to align UK defence expenditure with the average percentage spent on defence by the FRG, Italy and France', in Mary Kaldor, Dan Smith and Steve Vines (eds), *Democratic Socialism and the Cost of Defence*, Croom Helm, London, 1979, pp. 130, 136.

11 See J.C.M. Baynes, *The Soldier in Modern Society*, Eyre Methuen, London, 1972, pp. 45, 46.

following British practice, this is put at 15 per cent of the GDP per head. As it happens, this figure (and the assumptions behind it) is not one to which the calculations in the table are particularly sensitive. To take the case of France, if conscript pay were shifted from the assumed 15 per cent to 50 per cent of GDP per head, the figure in the final column would decrease from 4.0 to 3.9. No allowance, incidentally, is made for the gain experienced by the economy at large as a result of the net return of manpower to the civil sector implied - on this model - by the abolition of conscription, for the simple reason that it is too small materially to alter the results presented here. To take the case again of France, a net return of around 100,000 men to the civil sector would increase the size of the economically active population by less than 1 per cent and would accordingly increase GDP by only about the same amount.

But, finally, there are other costs associated with conscripts which are more important than their subsistence (in cash or in kind). The cost of training - measured perhaps in terms of the numbers of serving personnel within the professional core of the service allocated to the training role - is bound to be high. And because of the much higher throughput of conscripts compared to volunteers, whilst the latter also need to be trained, overall training costs are bound to be higher under a conscript regime. This will be true even after allowing for the possibility that our assumption of a 12-month training period might have to be modified as and when long-service volunteers need to undergo on-the-job training to familiarise themselves with any new types of weapons that enter service. Rather than attempt to quantify the training cost element, it can be allowed for by saying that the upward adjustments made in column 'j' to the figures given in column 'd' are *upper limits*, and almost certainly overstate the correction necessary to render NATO defence expenditure statistics compatible as between allies with different methods of manpower recruitment. Correspondingly, the 0.4 percentage point adjustment in the 1989 figure for the European NATO average percentage of GDP spent on defence (from 3.1 to 3.5 per cent according to the figures in Table 4.3, to allow for the relative cheapness of conscript forces) is also an upper limit.

Thus it would seem that even after allowing for differences in recruitment methods, to the end of the 1980s Britain continued to spend a significantly higher proportion of its GDP on defence than the NATO European average. The mean excess over the period since 1970 - even after making allowances for the conscript system practised by the European allies - was in the region of 1 percentage point. At 1989 levels, one percentage point comes to around one-quarter of the British defence budget or, in round figures, about £5 billion sterling (1989 pounds) annually.

Other special circumstances

There are two other possibilities that should not be overlooked which could explain the apparently excessive size of the British defence budget. One is the fact that Britain is a nuclear power; the cost of maintaining this status may have been added to the existing defence budget rather than replaced expenditure otherwise earmarked for conventional forces. Whilst, as already hinted, published figures relating to nuclear expenditure are not always trustworthy, and figures for total expenditure on the British nuclear weapons programme (i.e. including expenditure on tactical nuclear weapons) do not seem to be available, the cost of the nuclear programme would appear to be far too small to bridge the gap. In the financial year 1990/91, a year of peak expenditure on the new Trident programme, the cost of the strategic nuclear force was officially put at only 6 per cent of the defence budget.[12] In the more typical years of normal expenditure on nuclear weapons, the proportion was, not surprisingly, lower: in 1970,[13] 1.5 per cent; in 1975, 1 per cent; in 1980, 1.5 per cent; and in 1985, 2 per cent. If 3 per cent of the defence budget covers nuclear costs on average, it hardly begins to reduce the 25 per cent by which the British defence budget would seem to exceed the NATO European norm.

The second possibility is Ireland. Since 1970, British forces have been stationed and active in Northern Ireland at levels considerably in excess of what had been the case in the 1950s and 1960s. Again, however, the costs involved are too small to be of much interest here, even if we assume new expenditure has been involved, rather than a switching of resources from more obviously NATO-dedicated tasks. In 1989 the cost to the Ministry of Defence of British military involvement in Northern

---

12 For this figure, and others relating to nuclear expenditure except for the year 1970, see *Defence Statistics*, 1992 Edition, Government Statistical Service, HMSO, London, 1992, p. 9.

13 *Statement on the Defence Estimates*, 1970, Cmnd 4290, HMSO, London, 1970, p. 92.

Ireland (i.e. the cost appearing in the defence budget) was unofficially put at only £150 million a year;[14] even if this were an underestimate by 100 per cent, £300 million a year is only one and one-half per cent of the British defence budget.

It turns out that, for a satisfactory explanation of the size of British defence budgets it is necessary to return to first principles and to an analysis of the underlying factors determining the distribution of defence expenditure within alliances.

'Burden' sharing

It might be asked to what extent it is actually reasonable to expect alliance members to apportion their collective defence expenditure according to their share of the wealth-producing capacity of the alliance as a whole. One kind of answer assumes that each alliance member is facing the same risk of attack by a common enemy, much as a street of householders may share a common risk from flood or fire. Householders might choose to insure against the risk. If so, they would find that their premiums varied with the value of their houses - pretty well strictly proportionally were they all to insure with the same company. The analogy is imperfect, but if the GDP of a particular country is seen as the annual output derived from the capital and manpower assets of the country in question, and roughly proportional to the magnitude of these assets, defence expenditure in proportion to GDP can be seen as an insurance premium for the protection of these wealth-generating assets.

Equity criteria

Alternatively, rather than ask what might be 'rational', we can ask what might be 'fair'. How in equity should defence expenditure be distributed amongst a number of allies? The simplest answer might be for each ally to contribute to the common defence pool in proportion to what it could afford. The size of each member state's GDP is a measure of the scale of the country's wealth, so that a contribution - a defence budget - proportional to the size of GDP would seem fair.

---

[14] *Strategic Survey 1988-1989*, Brassey's for IISS, London, 1989, p. 106.

However there are other ways of looking at what equity would demand. One, due to de Strihou, focuses on differences between NATO members' ability to contribute to the common defence at the level of individual people in each country.[15] European NATO can be thought of, not as a collection of states in alliance, but as a number of regions of an imaginary pseudo-state with a defence budget to fund. As in any real state, according to this model the contribution of poorer regions to the cost of national defence will not be proportional to the total income of the region in question. Rather, we will see regions with a high average income per head contributing proportionally more than regions with poorer populations, because all governments levy personal taxes to fund common enterprises to a greater or lesser extent progressively with income. If we take GDP per head as standing for individual income, Table 4.3 shows that the British 'region' enjoys an average income only a little above the average of NATO Europe as a whole, whereas Germany's population (pre-unification) has an income about 50 per cent above the average. Thus if we supposed that the pseudo-state levied income tax to pay for defence (among other things) at a progressive rate similar to that applied by the British government to British taxpayers in the early part of 1979,[16] the proportion of income taken in tax would not be fixed, but would depend upon something like the cube root of the income in question[17]. Applied to our pseudo-state, the German 'region' of NATO would contribute about 3.5 per cent of its total income (GDP) to NATO Europe's defence costs, and Britain only 2.8 per cent of GDP.

Of course, in practice, the European NATO allies do not constitute a pseudo-state, and even if they were able at some stage to apportion defence expenditure among themselves in a manner that was collectively rational or 'fair' or both by some agreed criterion (equal proportions of GDP might be deemed both rational and fair), the actual sovereignty which each member state has over the size of its defence budget would

---

15 See Gavin Kennedy, *Burden Sharing in NATO*, Duckworth, London, 1979, pp. 60, 67.

16 Following, with a different base year and some simplifications of method, Kennedy, *op. cit.*, p. 62.

17 For UK income tax rates in early 1979, and a comparison with the less steep progression of tax take with increasing income in later years, see *The Economist*, March 14, 1992, p. 39. The cube root approximation holds reasonably well for incomes at and above the average level but overstates the tax taken from smaller incomes.

cause the distribution of defence expenditures within the alliance to adjust over time from the original apportionment to something else. (Here we are talking only about that part of defence budgets earmarked for alliance purposes. Some alliance members are or have been militarily involved in non-NATO activities which naturally also needs to be taken into account when explaining the overall size of defence budgets. This we turn to later.)

Free riding

There is some reason to believe that within functioning multi-member alliances there is an in-built tendency towards the exploitation by members making small contributions to the common defence of those members making large contributions. This is certainly believed to be the case for NATO in parts of the United States' government. A recent example of the resentment this can generate is found in a long document entitled 'Report on Allied Contributions to the Common Defense', submitted in May 1991 by the United States' Secretary of Defense to the US Congress. The production of the document was stimulated by another flare-up of Congressional suspicions that the NATO (and other) allies of the United States were not contributing their fair share to the costs of mutual defence, and the document was indeed required by legislation passed by the Congress between 1985 and 1989.

There are some theoretical indications that United States' suspicions may not be groundless. Alliances are undertakings by member states to contribute to a common pool of defence effort, for the good of all. Once created, with contributions apportioned according to some commonly agreed criteria (in practice it may simply be agreed, tacitly or otherwise, that states enter the alliance with the levels of defence expenditure already budgeted for the year in question), there will be a tendency at the sovereign state level for individual rationality to reassert itself. In other words there will be a tendency for all members of the alliance to reduce their contributions to the pool and also their share of their national income spent on defence as far as this is compatible with the continued survival of the alliance. There will be jockeying to try to shift the costs of the common defence onto other alliance members. Smaller members of the alliance will find this easier than larger members. For instance, the reduction in the sea power of the alliance as a whole caused by a member contributing, say, only one per cent of the total naval strength of the alliance deciding to cut its navy budget in half (and thus making considerable savings - perhaps fifteen or twenty per cent - on its own defence budget) will be in the region of only one half of one per cent. This will scarcely be noticeable when set against the natural fluctuations in the size of navies due to the normal operation of maintenance schedules.[18] Any attempt by the other members of the alliance to ensure that the state in question received only half the naval protection that it enjoyed before from its fellow members' navies would normally be impracticable.

This argument can readily be generalised from navies to a range of other sorts of provision for defence. In the case of Europe, Sweden and Switzerland have both benefited from the defences devised by NATO members to block (literally in some cases) Soviet expansion, but they contributed nothing at all to NATO.[19] There is no practicable way in which they could have been either denied these benefits or have been made to pay for them.

Existing alliance members could 'do a Sweden' in the sense of reducing their contribution to the alliance to small levels and still receive virtually the same benefits from it as they had before, provided they were small enough for the alliance, when deprived of part of their contribution, still to remain viable. So alliances are indeed vulnerable to 'drifting' towards a situation where smaller members contribute less than what might originally have been judged appropriate, and larger members more. In practice this will mean that an alliance that starts off with each member contributing, say, the same proportion of its GDP, will drift towards a situation where those members whose share of total alliance defence expenditure was larger than the average (because their GDPs were originally larger than the norm) finish up spending a higher proportion of their GDP on defence than the others. There is a sense in which high contributors to the common defence become locked into

---

[18] Strictly speaking such precision about the consequences of a reduction in ships of one half of one percent is unjustified. The capability of a navy of a certain size may be linear with size (once allowance is made for the diversity of ship types), but it may equally well depend on the number of ships in the fleet in some other way, for instance with the square. In the latter case a reduction of one half of one percent in fleet size would reduce the fighting strength of the alliance by one percent - still of course normally small enough not to be noticed.

[19] By 1989 Sweden was spending 2.4% of its GNP on defence, Switzerland 1.8%. The NATO European average was 3%. None of these figures allows for the understatement of true defence costs normally implied by conscripted forces.

the position of being members of good standing because of the difficulty of reducing their contribution to the common defence without threatening the future of the alliance as a whole. If it should also happen to be the case that the GDPs of high contributors do not grow over time as rapidly as those of other alliance members, the drift away from the original (or otherwise deemed desirable) equal proportions of GDP is further accelerated. In 1962, for instance, Britain had the highest share of the European NATO defence budget and the highest GDP in Europe. In 1989 it still had the highest share of the European NATO defence budget (along with France), but its GDP had slipped to fourth in the European league.

The above informal analysis is broadly similar but not identical to the very well known theory of intra-alliance defence expenditures due to Olson and Zeckhauser.[20] It is sufficiently different in a number of respects to require further analytical support and this is provided in the second appendix to this Annex.

If the foregoing is accurate, we should expect some statistical support from data on NATO defence expenditures. We should expect to find that, broadly speaking, member states with high shares of the total defence budget of the alliance pay a higher proportion of their GDP towards defence than member states with small shares of the total defence budget.

Statistical tests using regression analysis were carried out for NATO patterns of defence expenditure for the years 1970, 1975, 1980, 1985 and 1989, with the results (as well as data) presented in the Tables in Appendix 1.

Each test shows a statistically significant relationship between the proportion of national expenditure allocated to defence (see Appendix 1 for details) and the absolute size of the defence budget of the country concerned. The test in fact is one of correspondence between the proportion of GDP spent on defence and the logarithm of the defence budget. The latter gives a stronger correlation than the crude defence budget figure; the reason for this may be connected with the possibility alluded to in Appendix 2 that non-linear elements may be present in the relationship between the two.

The conclusion of this first statistical test is a simple one; namely that

---

[20] M. Olson and R. Zeckhauser, 'An Economic Theory of Alliances', *Review of Economics and Statistics*, 48 (3), (1966), pp. 266, 279.

British defence expenditure is as high as it is partly because Britain was originally a major contributor to the NATO alliance and, like other major contributors, has found it relatively difficult to reduce its contribution for fear that this might affect the stability of the alliance as a whole. Whilst this does not rule out the possibility that Britain too may have been able to reduce defence expenditure somewhat and acted as a 'free rider' on the contribution made by the United States, it concludes that Britain has 'free ridden' less than it has been 'free ridden' on.

Parallel studies

That NATO is prone to free-riding tendencies which may be endorsed by statistical analysis is not by any means a novel suggestion. A pioneering statistical study by the defence economist (and polymath) Gavin Kennedy in 1975 - upon which the present analysis is to some extent modelled - found some weak evidence of free-riding behaviour within NATO.[21] He also found that this tendency seemed to be diminishing with time. However, Kennedy, following (albeit with a critical eye) the original Olson and Zeckhauser analysis, was looking for correlations between the share of GDP spent on defence and total GDP of the state in question, rather than the size of its defence budget, which is what the analysis presented here suggests may be more appropriate.

Other analysts have also looked at the question. In some respects a statistically more promising technique than the one employed here relies on figures for the variation in British defence expenditure over a long period of time and aims to explain these differences according to a comparatively wide list of possible explanatory variables, including the defence expenditure undertaken by allies. One such study, by Hartley and McLean,[22] finds some slight evidence of free riding by Britain on allies' defence expenditures. Another, by Hansen, Murdoch and Sandler,[23] finds evidence of free riding by Britain (and France), but with

21 Gavin Kennedy, *op. cit.*

22 K. Hartley and P. McLean, 'UK Defence Expenditure', *Public Finance* 36, 2, 1981, pp. 171, 192.

23 Laura Hansen, James C. Murdoch and Todd Sandler, 'On Distinguishing the Behavior of Nuclear and Non-Nuclear Allies in NATO', *Defence Economics*, Vol. 1, No. 1, 1990, pp. 37, 55.

respect only to expenditure on nuclear forces. That some sorts of defence activity by allies are more easily taken advantage of by would-be free riders than others would seem natural. A slightly different type of analysis of the correlates of the level of British defence expenditure over time, by Smith,[24] also uncovers statistical evidence of free riding by Britain on United States' defence expenditure.

What is novel about the analysis presented here is the emphasis it gives to the possibility, not just that free riding occurs in NATO, nor even that Britain may have free ridden on the United States, but that Britain itself has been free ridden upon. Put another way, the British share of GDP spent on defence is higher than the NATO European average partly because Britain - to repeat - has been more free ridden upon than it has free ridden.

Out of area

The statistical analysis shows that if the free-riding tendencies of smaller allies on the backs of larger were the only factor at work in determining the distribution of defence spending as between NATO allies, about one-third of the difference between Britain's proportion of GDP being spent on defence and the European NATO average figure could be accounted for. The column at the extreme right of the Tables 4.4-4.8 estimates what the British and other national figures for defence expenditure as a proportion of GDP would be if free-riding were the only factor at work.

However, an apparently much more complete explanation of how defence expenditures vary as between NATO allies can be obtained if the additional factor of differences in the extent to which NATO members have pursued military objectives outside the geographical and strategic boundaries of the NATO alliance is also taken into account.

Dummy variable

The statistical technique employed to incorporate this factor smoothly into the analysis requires each NATO member to be allocated a binary - yes/no - variable according to whether they either pursue or do not

---

[24] R. P. Smith, 'The Demand for Military Expenditure', *Economic Journal*, 90, 1980, pp. 811, 820.

pursue (for the year in question) military objectives outside the NATO area. This 'dummy variable' technique is not perfectly satisfactory since it takes no account of differences in the intensity with which such objectives may have been pursued. But it does at least avoid the difficulties that would be encountered when attempting to place precise figures upon, for instance, the degree to which the end of the Vietnam war coincided with a corresponding increase in the proportion of US defence spending orientated towards NATO.

Table 4.2 shows particularly high United States' and Portuguese figures for the defence share of GDP during the period 1970/75, partly accounted for by the fact that during this period each country was engaged in a war outside the NATO area. Similarly, both Greece and Turkey display high figures between 1975 and 1985, which coincided with their being on something approaching a mutual war footing during this period over their quarrels concerning the future of Cyprus after the fall of President Makarios. The dummy variable is capable of registering only very large shifts. So Portugal is awarded a score of 1 for military activities out of the NATO area for the years 1970 and 1975 (during or just after its colonial wars in Africa), but zero for the other years analysed. The USA retains a score of 1 throughout to reflect its acknowledged global pattern of military alliances. Greece and Turkey are also given a consistent score of 1, partly to reflect their unresolved quarrels over the future of Cyprus and partly to reflect other rivalries between them in the Aegean.

Britain and France are also awarded a score of 1 throughout. France's score reflects its undisguised interest in remaining a military power of some account out of the NATO area, particularly in Africa. Britain's score reflects the supposition - to which this book aims to give credence - that France is not alone and that Britain too has continued, at the operational level, to pursue a 'global policy' well after 1970, whatever declaratory policy seems to have said.

The addition of the dummy variable to the statistical analysis significantly improves the free-rider explanation for differences in defence budgets as a proportion of GDP as between NATO allies. While the free-rider explanation accounted for approximately 40 per cent of the variation, the out-of-area factor explains a further 30 or 40 per cent.

Conclusion

Statistical analysis rarely permits any conclusion to be drawn in which one can have absolute confidence. The conclusions of the analyses undertaken here are presented in descending order in terms of the degree of confidence that ought to be invested in them.

The first is that, since 1970, Britain has without doubt devoted a higher proportion of its GDP to defence than the European NATO average. This extra amount has never been less than about one percentage point, i.e. Britain has consistently spent between 1 and 1.5 per cent more of its GDP on defence than its European allies. There is, admittedly, some uncertainty concerning the accuracy and comparability of the published data, but these uncertainties are very much smaller than the apparent excess.

Secondly, the more obvious explanations for such an excess - the fact that, unlike most of its European allies, Britain relies on professional, all-volunteer forces and is both a nuclear power and engaged in a minor but persistent internal conflict (Ireland) - are not convincing, because the costs that can plausibly be associated with these distinctive factors and that feature in the defence budget are simply too small. The maximum extra costs associated with volunteer forces could possibly add four-tenths of a percentage point to the British share of GDP spent on defence. This brings the net typical difference over the twenty-year period since 1970 between the British figure and the European NATO average down to about one percentage point.

Thirdly, a parsimonious single explanation for the one percentage point excess does not seem to be available. But two explanations taken together may be capable of accounting for it. One of these arises because NATO, like any similar alliance, suffers from a tendency for smaller contributors to the common defence to cut back even on what little they do contribute, in the usually justified expectation that their bigger allies will most likely tolerate, or at most merely grumble about, their underpayment than do anything about it. Thus it is not so much that Britain pays too great a share of the common defence, but that its smaller allies (smaller in the sense that their original contributions to the common defence were smaller than that of Britain) have learnt how to pay less and less. The other explanation is a good deal more straightforward and relies on the proposition that, unlike most of its European allies, Britain remains militarily involved in non-NATO business or, in shorthand, out of area. There is no obvious way in which the additional costs of an out-of-area capability can be directly calculated (but see Section Two and the Conclusion of the main text), especially when it may take the form, not of extra men or machines, but of giving both (at a cost) a multipurpose capability in order to render them fit for employment outside as well as inside the NATO area. But statistical analysis suggests that out-of-area costs may be large. Of the one per cent of the British GDP excess (once differences between the cost of volunteer and conscript forces have been allowed for), three-fifths is accounted for by out-of-area activities. In money terms, for 1989, the sum involved is about £3 billion.

Table 4.1[25]

National NATO Defence Expenditures as Percentages of GDP According to Four Authorities: 1989

	NATO	IISS	Pentagon	SIPRI
Belgium	2.5	2.5	2.6	2.5
Denmark	2.1	2.1	2.1	2.1
France	3.7	3.7	3.7	3.7
Germany	2.8	2.8	2.8	2.8
Greece	5.7	5.8	5.7	6.8
Italy	2.3	2.5	2.4	2.4
Luxembourg	1.1	1.1	1.2	1.0
Holland	2.9	2.9	2.8	2.9
Norway	3.2	3.3	3.2	3.2
Portugal	3.2	3.1	3.2	3.9
Spain	2.0	2.1	2.0	2.1
Turkey	4.3	3.7	4.0	3.9
Britain	4.1	4.2	4.1	4.0
Eur. NATO Average	3.1	3.1	3.1	
Canada	2.0	1.9	2.0	2.0
USA	5.9	5.8	5.9	5.8

[25] Sources for the figures are as follows:
NATO: *Financial and Economic Data Relating to NATO Defence*, NATO Press Service, Release M-DPC-2(91) 105, Brussels, 12 December 1991.
IISS: *The Military Balance 1990-1991*, Brassey's for the International Institute for Strategic Studies, London, 1990, p. 17 and pp. 59, 83.
Pentagon: *Report on Allied Contributions to the Common Defense*, May 1991. A Report to the United States Congress by the Secretary of Defense, p. A-14 and *passim*.
SIPRI: *SIPRI Yearbook 1991: World Armaments and Disarmament*, OUP, Oxford, 1991, pp. 174, 180.

Table 4.2[26]

Defence Expenditures as % of Gross Domestic Product

	Av. 70/74	Av. 75/79	Av. 80/84	Av. 85/89	1987	1988	1989	1990
Bel	2.9	3.2	3.3	2.9	3.0	2.7	2.5	2.4
Den	2.4	2.4	2.4	2.1	2.1	2.1	2.1	2.0
Fra	3.9	3.8	4.1	3.8	3.9	3.8	3.7	3.6
Ger	3.5	3.4	3.3	3.0	3.1	2.9	2.8	2.8
Gr	4.7	6.7	6.6	6.3	6.3	6.3	5.7	5.8
It	2.5	2.0	2.1	2.3	2.3	2.3	2.3	2.1
Lux	0.8	1.0	1.2	1.1	1.2	1.3	1.1	1.1
Hol	3.1	3.2	3.2	3.0	3.1	3.0	2.9	2.7
Nor	3.3	3.1	2.9	3.2	3.3	3.2	3.2	3.2
Por	6.9	3.9	3.4	3.2	3.1	3.2	3.2	3.1
Spa			2.3	2.2	2.4	2.1	2.0	1.8
Tur	4.4	5.7	4.9	4.3	4.2	3.8	4.3	4.9
UK	5.0	4.9	5.2	4.6	4.6	4.2	4.1	4.0
I	3.6	3.7	3.6	3.3	3.3	3.2	3.1	3.0
Can	2.1	1.9	2.1	2.1	2.1	2.1	2.0	2.0
USA	6.5	5.1	5.8	6.3	6.5	6.1	5.9	5.7
II	6.1	4.8	5.5	6.0	6.1	5.7	5.5	5.3
III	5.2	4.6	4.7	4.8	4.8	4.5	4.4	4.1

I) European NATO average.
II) North American NATO average
III) Overall average.

---

[26] 'Financial and Economic Data Relating to NATO Defence', NATO Press Release M-DPC-2(91) 105, December 1991. Iceland, having no armed forces, is not included. Germany's expenditures related to Berlin not included. The figures for the NATO European average for the periods 1970-74 and 1975-79 are the author's estimates based upon entries in issues of *The Military Balance*, IISS, London between the years 1973-74 and 1979-80.

Table 4.3[27]

Defence/GDP Standardised for Different Manpower Recruitment Methods: NATO 1989

	a	b	c	d	e	f	g	h	i	j
Bel	9.9	3.9	155	2.5	92	36	12	18	4.4	2.8
Can	26.6	10.5	543	1.9	90	0			10.5	1.9
Den	5.1	2.2	105	2.1	32	11	12	6	2.4	2.3
Fra	56.4	35.1	958	3.7	461	235	12	118	38.5	4.1
Ger	60.4	33.9	1210	2.8	469	204	15	123	38.2	3.2
Gr	10.1	3.2	55	5.8	162	127	20	89	4.1	7.4
Ic	0.2	0	5	0	0	0			0	0
It	57.3	21.7	858	2.5	390	259	12	129	26	2.9
Lux	0.4	0.1	7	1.1	0.8	0			0.1	1.1
Hol	14.8	6.4	222	2.9	103	46	16	29	7.2	3.2
Nor	4.2	3.1	93	3.3	34	22	12	11	3.5	3.8
Por	10.5	1.3	42	3.1	68	43	15	26	1.5	3.6
Sp	39.9	7.7	373	2.1	274	201	12	100	9.3	2.5
Tur	55.9	2.8	79	3.7	647	554	18	369	3.9	4.8
UK	56.6	34.9	839	4.2	306	0			34.9	4.2
US	249	300.3	5199	5.8	2118	0			300.3	5.8

a) population of member state, in millions.
b) defence expenditure in billion (10^9) 1989 $US.
c) gross domestic product, billion 1989 $US.
d) ratio (b):(c) percent. e) active strength of the armed forces, in thousands
f) conscripts (thousands) included in the armed strength.
g) period of (army) conscription, months.
h) additional number (thousands) of volunteers theoretically necessary (see above), to replace the given number of conscripts.
i) estimated size of the defence budget, billions 1989 US dollars, with conscripts replaced by the appropriate number of volunteers. j) ratio (i):(c) percent.

27 Defence expenditure figures in this table are taken from the IISS version of NATO standard defence expenditures. These figures and other data used in the table are given in *The Military Balance 1990-1991*, Brassey's for IISS, London, 1990, p. 17 and pp. 59, 86.

Appendix 1: Regression Analysis

This appendix contains details of the regression analysis carried out on five years of data relating to the distribution of the 'defence burden' as between the NATO allies. What is sought is an explanation for the proportion of GDP spent on defence and how this varies as between the allies. This is accomplished in two stages. The first explanation tested is similar to that of Olson and Zeckhauser (see main text in this Annex), and would account for the variations in percentage of GDP spent in terms of 'free riding', or the exploitation by smaller allies of their larger partners. Small defence:GDP ratios should follow from small defence budgets. In fact, the statistical test employs the logarithm (to the base ten) of the defence budget (unadjusted for differences in manpower recruitment methods). The justification for taking the logarithm is primarily that it works, in the sense that, once incorporated into the statistical analysis, it fits the facts better than the simple defence budget figure does. Another justification comes from the analysis given in the second appendix of 'free riding', where a non-linear relationship between defence budget:GDP ratios and defence budgets is held out at least as a possibility on theoretical grounds.

The suggested relationship between the defence budget:GDP ratios and defence budgets as derived from the statistical regression analysis is given by Regression Equation ONE. How well this fits the actual facts is given by the R squared (R sqd) value. A perfect fit would produce an R squared of 1. If we treat this equation as perfectly accurate, we can produce a 'predicted' defence budget:GDP ratio (by inserting the appropriate defence budget from the Def Bud column into the equation). This generates the column of figures at the extreme right-hand side of the Tables (P1). It can be seen that these predictions are somewhat adrift from the actual figures given in the fourth (DB/GDP) column.

The second stage is to incorporate an additional explanation in the form of non-NATO military activities. Where these are significant, a score of 1 is entered in the third (out of area - ooa) column of figures: where insignificant, 0 is entered. Higher defence budget:GDP ratios should follow higher levels of non-NATO military activity. A new regression analysis is then carried out and the suggested relationship between what are now three factors is given in Regression Equation TWO. The higher R squared in every case indicates that the new explanation is more complete than the old, and this is confirmed by the new column of predictions (sixth column of the Tables - P2) which are now much closer to the actual (fourth column) figures.

The predictions are in an important sense estimates only and ought properly to be stated, not in the form of single figures, but as a bracketed range. The numbers written beneath each regression equation, when added to the numbers immediately above, produce a new equation determining the upper limit to the predicted figure. Subtracting the lower tier of numbers produces the equation for the lower limit.

Analysis was restricted to five well-spaced years, essentially chosen at random (i.e. there is no reason why decade end or middle years should see untypical levels of military expenditure or activity). It would have been possible to carry out a separate analysis for every year of the period under study, but this would not have added very much that is new. Defence budgets change slowly and the size of a budget in any one year owes something (at least in the statistical sense) to its size the previous year.

For reasons of consistency, where possible, all data were taken from the appropriate editions of the IISS's *Military Balance*. Where there were gaps in non-defence data, OECD figures were employed. All money figures are in billions of United States dollars.

The figure for the defence budget for Iceland is not taken to be zero (which is the official figure) but a nominal $10,000. This avoids many technical computational difficulties without altering the basis of the analysis in any way. For presentational reasons, some rounding of raw figures has proved necessary. All computation was performed with unrounded figures.

Table 4.4

1970

Country	Def Bud	log DB	ooa	DB/GDP	GDP	P2	P1
Belg	0.68	-0.16749	0	2.6	26.1	2.7	3.5
Can	1.98	0.296665	0	2.4	82.6	3	3.9
Den	0.36	-0.4437	0	2.4	15	2.5	3.2
Fra	4.9	0.690196	1	3.3	148.6	5.1	4.2
FRG	6.23	0.794488	0	3.3	188.8	3.3	4.3
Gr	0.45	-0.34679	1	4.7	9.5	4.5	3.3
Ic	0.00001	-5	0	0.002	0.47	-0.3	-0.6
It	2.59	0.4133	0	2.8	92.6	3.1	4
Lux	0.008	-2.09691	0	0.8	1	1.5	1.8
Holl	1.1	0.041393	0	3.5	31.4	2.8	3.7
Nor	0.37	-0.4318	0	3.3	11.3	2.5	3.3
Por	0.36	-0.4437	1	5.6	6.4	4.4	3.2
Tur	0.56	-0.25181	1	3.6	15.5	4.5	3.4
UK	5.95	0.774517	1	4.9	121.5	5.2	4.3
US	72.25	1.858838	1	7.4	976.4	5.9	5.2

NATO European Average DB/GDP = 3.5%
Regression Equations (DB/GDP written as D/G)

ONE	D/G	=	3.6	+	0.85	log DB			
			1.3		0.22				
						R sqd	=	0.54	
TWO	D/G	=	2.8	+	0.64	log DB	+	1.9	OOA
			0.9		0.17			0.5	
						R sqd	=	0.78	

Table 4.5

1975

Country	Def Bud	log DB	ooa	DB/GDP	GDP	P2	P1
Belg	1.97	0.294466	0	3	66.4	2.7	3.9
Can	2.96	0.471292	0	2.2	137	2.8	4
Den	0.94	-0.02687	0	2.2	41.9	2.5	3.6
Fra	14	1.146128	1	3.9	359.2	5.9	4.5
FRG	16.1	1.206826	0	3.6	441.6	3.1	4.6
Gr	1.43	0.155336	1	6.9	20.8	5.4	3.8
Ic	0.00001	-5	0	0.0008	1.25	0.2	-0.3
It	4.7	0.672098	0	2.6	177.5	2.9	4.2
Lux	0.02	-1.69897	0	0.9	2.2	1.8	2.3
Holl	2.98	0.474216	0	3.6	83.4	2.8	4
Nor	0.93	-0.03152	0	3.1	29.6	2.5	3.6
Por	1.09	0.037426	1	6	18.1	5.4	3.7
Tur	2.2	0.342423	1	6.6	33.1	5.5	3.9
UK	11.1	1.045323	1	4.9	226	5.8	4.5
US	89	1.94939	1	5.9	1499	6.2	5.2

NATO European Average DB/GDP = 3.8%
Regression Equations

ONE	D/G	=	3.6	+	0.79	log DB				
			1.7		0.27					
						R sqd	=	0.39		
TWO	D/G	=	2.5	+	0.46	log DB	+	2.8	OOA	
			1.0		0.17			0.6		
						R sqd	=	0.80		

Table 4.6

1980

Country	Def Bud	log DB	ooa	DB/GDP	GDP	P2	P1
Belg	3.73	0.571709	0	3.3	113	2.8	3.4
Can	4.24	0.627366	0	1.7	249.4	2.8	3.5
Den	1.4	0.146128	0	2.4	58.3	2.6	3.1
Fra	20.22	1.305781	1	3.9	518.5	4.9	3.9
FRG	25.12	1.40002	0	3.2	785	3.2	4
Gr	1.77	0.247973	1	5.1	34.7	4.3	3.2
Ic	0.00001	-5	0	0.0004	2.8	0	-0.4
It	6.58	0.818226	0	2.4	274.2	2.9	3.6
Lux	0.05	-1.30103	0	1	5	1.8	2.1
Holl	5.24	0.719331	0	3.4	154.1	2.9	3.5
Nor	1.57	0.1959	0	2.9	54.1	2.6	3.2
Por	0.89	-0.05061	0	3.8	23.4	2.5	3
Tur	2.92	0.465383	1	4.2	69.5	4.4	3.3
UK	24.45	1.388279	1	5.1	479.4	4.9	4
US	143	2.155336	1	5.5	2600	5.3	4.5

NATO European average DB/GDP = 3.6%
Regression Equations

ONE	D/G	=	3.0	+	0.69	log DB				
			1.1		0.17					
						R sqd	=	0.56		
TWO	D/G	=	2.5	+	0.51	log DB	+	1.7		OOA
			0.7		0.13			0.40		
						R sqd	=	0.81		

Table 4.7

1985

Country	Def Bud	log DB	ooa	DB/GDP	GDP	P2	P1
Belg	2.69	0.429752	0	3.3	80.34	2.7	3.5
Can	7.4	0.869232	0	2.2	340.9	2.9	3.9
Den	1.3	0.113943	0	2.3	57.12	2.6	3.3
Fra	20.73	1.316599	1	4.1	511.4	5.6	4.2
FRG	19.92	1.299289	0	3.2	621.7	3.2	4.2
Gr	2.33	0.367356	1	7.1	32.65	5.2	3.5
Ic	0.00001	-5	0	0.0004	2.7	0	-0.6
It	9.46	0.975891	0	2.6	360.6	3	4
Lux	0.04	-1.39794	0	1.2	3.34	1.8	2.1
Holl	3.98	0.599883	0	3.2	124.2	2.8	3.7
Nor	1.79	0.252853	0	3.1	57.91	2.6	3.4
Por	0.63	-0.20066	0	3	20.79	2.4	3
Sp	4.8	0.681241	0	2.8	169.1	2.8	3.7
Tur	2.29	0.359835	1	4.4	52.3	5.2	3.5
UK	25.36	1.404149	1	5.3	480.6	5.7	4.3
US	266.6	2.42586	1	6.9	3839	6.2	5.1

NATO European Average DB/GDP = 3.7%
Regression Equations

ONE	D/G	=	3.2	+	0.77	log DB			
			1.4		0.22				
						R sqd	=	0.47	
TWO	D/G	=	2.5	+	0.50	log DB	+	2.5	OOA
			0.9		0.15			0.5	
						R sqd	=	0.81	

Table 4.8

1989

Country	Def Bud	log DB	ooa	DB/GDP	GDP	P2	P1
Belg	3.9	0.591065	0	2.5	155.2	2.4	3
Can	10.55	1.023252	0	1.9	542.6	2.6	3.3
Den	2.2	0.342423	0	2.1	105.1	2.3	2.9
Fra	35.1	1.545307	1	3.7	958.1	4.7	3.6
FRG	33.9	1.5302	0	2.8	1210	2.8	3.6
Gr	3.2	0.50515	1	5.8	55.5	4.3	3
Ic	0.00001	-5	0	0.0002	5.1	0.1	-0.4
It	21.7	1.33646	0	2.5	858	2.7	3.5
Lux	0.08	-1.09691	0	1.1	7.4	1.7	2
Holl	6.4	0.80618	0	2.9	222	2.5	3.2
Nor	3.1	0.491362	0	3.3	92.7	2.4	3
Por	1.32	0.120574	0	3.1	42.1	2.2	2.8
Sp	7.75	0.889302	0	2.1	373	2.5	3.2
Tur	2.88	0.459392	1	3.7	78.7	4.3	3
UK	34.9	1.542825	1	4.2	838.6	4.7	3.6
US	300.3	2.477555	1	5.8	5199	5.1	4.2

NATO European Average DB/GDP = 3.1%

Regression Equations

ONE	D/G	=	2.7	+	0.62	log DB				
			1.1		0.18					
						R sqd	=	0.46		
TWO	D/G	=	2.2	+	0.42	log DB	+	1.9	OOA	
			0.7		0.12			0.4		
						R sqd	=	0.78		

Appendix 2: Game Theory

The reference above to the failure of rational action on the part of the individual alliance member to correspond to what may seem rational for the whole carries echoes of the famous prisoner's dilemma; indeed, game theory proves to be an appropriate tool to carry the analysis forward in this case. The game is still that of the prisoner's dilemma, but in the present case the number of players in the game will normally be more than two.

On the left-hand side of Table 4.9 we have one alliance member, Erehwon, out of the total number of closely similar states, n, looking at the consequences of its choices with regard to either remaining an alliance member of good standing, at a cost to it of C, or alternatively saving money by cutting its defence budget (this is assumed, for sake of simplicity, to be a one hundred per cent cut in this case). We are thinking of member states as having formed an alliance, and analysing how stable the alliance arrangement may be.

In order to determine along which path rational action lies, Erehwon will first calculate what its payoff would be if it continued to maintain its defence budget at the original level. It cannot be sure that the others will also continue to keep their defence budgets high, but can ascribe a probability, p, that if it, Erehwon, stays in good standing, then the others will too (there is a simplification here that sees the other (n-1) parties to the arrangement as all having the same probability, p, of behaving as described; in reality this will not be exactly true since we have described the states as being closely similar rather than identical).

Table 4.9
The Alliance Member's Dilemma

		Others	
		Maintain	Cut
Erehwon	Maintains	B - C	B/n - C
	Cuts	(n - 1)B/n	0

Correspondingly, there is a probability (1-p) that, even if Erehwon maintains its defence budget, the other members of the alliance will decide not to do so, leaving Erehwon much less well defended than before. How much less well defended is not perfectly clear: B/n is indicative rather than anything else of what the new defence benefits to Erehwon would be if it were deserted by its (n-1) allies compared to the B it would obtain within a solid alliance. So Erehwon's payoff from a choice to maintain its defence budget will be:

$$p(B-C) + (1-p)(B/n - C)$$

Erehwon will also calculate the payoff from the alternative of cutting its defence budget, bearing in mind that there will now be a new probability, q, that the others will nonetheless keep their defence budgets high. Erehwon can be confident that q will be less than p; probably also that, the larger the number of players Erehwon is one of, the smaller the difference is likely to be between p and q, since the defection of one of a large number is less likely to lead to a dissolution of the whole arrangement than the defection of one in a setting of a small number of players.

Erehwon's payoff from deciding to cut its defence budget is:

$$q(n-1)B/n$$

In terms of this analysis, Erehwon will continue to maintain its defence budget if the difference between the 'maintains' payoff and the 'cuts' payoff is positive - that is to say if it pays Erehwon to do so. From the above, this difference, D, can be shown to be:

$$B[1 + (n-1)(p-q)]/n - C$$

We can see by inspection of the above expression for D that where n is large, D becomes negative (recalling that (p-q) is of the order of 1/n). That is to say, there is no incentive for a player in a large alliance to maintain its defence budget at a high level and that therefore the stability of a collective arrangement for mutual defence of the size of NATO would seem to be doubtful since here we are dealing with twenty or so players.

At this point a word of caution is necessary. We have so far assumed that B is independent of n which may not be entirely justified.

Non-linear effects

The rationale for the simplest imaginable alliance (one between two identical units) is that both allies can benefit from the pooling of their defence resources (armies, navies and air forces plus defence-industrial and manpower potentials) *and* of their defence liabilities (territory etc. to be defended). It seems hardly likely that alliances would be formed as frequently as is the case if, as in this example, a doubling of defence capabilities were exactly matched by a doubling of liabilities. In fact there is some reason to think that there may be economies of scale in translating defence resources into defence capabilities. The British pioneer of operational analysis, F. W. Lanchester, famously contended that the fighting strength of an armed force depended on the *square* of its size. Thus an alliance of two equal units would have twice the defence liabilities of each, but four times the fighting strength. Lanchester's claims have been extensively disputed and Lanchester himself did not consider that his 'square law' should replace the common-sense linear (ordinary addition of fighting strengths) calculation in all circumstances.

If Lanchester's square law did apply without modification to the above game-theory analysis of how the benefits and costs of belonging to an alliance of n members impact on an individual member, it can be shown that the difference between costs and benefits will not depend on the size of the alliance at all, and that a large alliance will be as stable as a small one. If, almost certainly more realistically, there were both linear and square terms involved in the assessment of the size of the defence capabilities available to an alliance on the pooling of resources, we would still expect large alliances to be less stable than small, but more stable than the simple game-theory analysis above appears to suggest.

In reality there will be other checks to instability in the form of bilateral connections between large alliance members and small, where tendencies by the latter to economise at their allies' expense may be held at bay, for example, by preferential arrangements for the transfer of armaments.

Small is powerful

Returning to the simple theory, we can take the analysis a little further with a new assumption that the players in the game are not all of the same size, only that nearly all of them are, with one player being the odd one out, either much larger than the others or much smaller. We can call this player Ytiddo. It differs from the rest in that, whilst all the other players apart from Ytiddo contribute to the collective defence at a level close to the average, Ytiddo contributes at a rate 'e' times the average. The size of e could be less than one for an exceptionally small contributor, or greater than one for an exceptionally large contributor. The latter, for instance, could be a large contributor because its GDP at the time the alliance was formed was significantly larger than the average.

From Table 4.10 (see also the Coda to this appendix) the difference D for Ytiddo between maintaining its defence budget and cutting it is:

$$B[e + (p^{\sim} - q^{\sim})(n-e)]/n - C$$

Again we can see that increasing n causes the value of the above expression to become negative, with what was an incentive to maintain the size of the defence budget becoming a disincentive. We also notice something else. The above expression for D is identical to the earlier expression for D if we replace n by n/e. In other words the dilemma facing the exceptionally large (say 4 times the norm) player is the same as that facing any player in a uniform group whose size has been reduced by a factor 4. Also the dilemma facing a small player (tenth of the norm) is the same as that facing a player in a uniform group ten times the size. Put another way, in a grouping of a given size, the temptation for a larger than average player to defect will be smaller than average, and the temptation for a smaller player, larger than average. Correspondingly the term $p^{\sim}-q^{\sim}$ depends not solely on n, but on e too, in that a very much smaller than average player would be less likely to induce retaliation for its defection than would an average sized player or *a fortiori* an above average sized player. Accordingly $p^{\sim}-q^{\sim}$ will have a value of the order of e/n.

Not all alliances are equally vulnerable to what it is customary in discussions of this sort to describe as 'free riding' or 'easy riding'; that is to say, the tendency of alliance members to exploit their larger partners. A small alliance with three or four members all making about the same (one-third or one-quarter) contribution to the military strength of the collective would be less prone to this sort of behaviour than a large alliance, whose members moreover varied greatly in the original contribution they made to the collective. NATO falls into the latter category, and it is beginning to seem as if the complaints emanating from the United States about failures of its NATO allies to pay their fair share to the common defence have some basis.

Table 4.10
The Odd One Out

		Others	
		Maintain	Cut
Ytiddo	Maintains	B - C	B.e/n - C
	Cuts	(n - e)B/n	0

Coda

Suppose we had four players.

	Others			
	all coop	2 coop	1 coop	0 coop
coops	$p^3(B-C)$	$3p^2(1-p)(B-s-C)$	$3p(1-p)^2(B-2s-C)$	$(B-3s-C)(1-p)^3$
Ytiddo				
defects	$q^3(B-e.s')$	$3q^2(1-q)(B-e.s'-s)$	$3q(1-q)^2(B-e.s'-2s)$	0

The cooperative, defence budget-maintaining, payoff (CPO) for Ytiddo is:

$$B - C - s(3-3p)$$

And the non-cooperative, defence budget-cutting, payoff (NCPO) for Ytiddo is:

$$3qs$$

In general, for n players, CPO - NCPO is given by:

$$B - C + (n-1)(p-q)s - (n-1)s$$

By definition, in general:

$$e.s' + (n-1)s = B$$

$$s' = B/n$$

so

$$s = [B - e.B/n]/(n-1)$$

And CPO - NCPO becomes:

$$B[e + (p-q)(n-e)]/n - C$$

Index

Abbot artillery piece 48
ACOST 120, 145, 149
Action Plan
of IEPG 98, 103
Aden
Britain's withdrawal from 73
Aircraft carriers 63, 65, 80, 82
ALARM air-to-surface missile 104-106, 137, 139-140
Alvey 149
Antelope HMS 79
Ardent HMS 79
Argentina 84, 162
Ark Royal HMS 64, 81
Armilla patrol 75
Arms exports 146
as catalyst for other exports 142
to Arabian peninsula 162
Asteris, M. 104
AWAC aircraft
distress purchase of 103

Baker, R. H. ix
Ball, Desmond 1, 5, 8, 32, 39, 166
Baylis, John 35, 37
Beach, General Sir Hugh 47, 56
Bell, Coral 38, 39
Bellamy, Christopher 50
Bittleston, Martyn 98, 113, 115, 116, 121, 130, 148
British Aerospace 17, 99, 105, 121, 122, 130, 136, 138, 139, 142
Brunei 67, 72, 83
British military presence in 76-77
financial assets 73
Bryson, Vice Admiral 126
Bulwark HMS 81
'Buy British' 90, 93, 96, 98-100, 124
motto of development industrial circle 89
premium involved 124-129, 160-162
Byrd, Peter 26, 140

Cabinet Office 154, 155

Cable, Sir James 37, 38
Callaghan, James 78
CENTO 67, 68, 75
Challenger main battle tank 50, 106-107, 126
Chalmers, Malcolm 31, 35
Chandler, Sir Colin 107, 138
Chevaline 21, 108, 169
Chieftain main battle tank 48, 140
Iran order for 106, 126
CIA
estimates of British arms exports 119
Circles of social actors 13, 26, 162
declaratory versus operational and industrial 165
Coded language 4, 45, 63, 80
used by RN to disguise true nature of their ships 80-82
Coker, Christopher 37, 49, 68
Conservative government 157
and out-of-area 69-71
Conventional wisdom of British defence thinking 36, 38
Cooper, Sir Frank 117
County Class light cruisers 80
Coventry HMS 79
Cruise missiles 79
Cyprus 67, 70, 182
British military presence in 76-77

Declaratory/development-industrial inconsistencies 89, 135-152, 160
and the Procurement Executive 144
Declaratory level of policy 2, 5, 14-22, 39, 41, 64, 68, 78-80, 88, 90-98, 162, 167
Declaratory/operational inconsistencies 66-84, 88, 131, 160
and flexible response 60-62
and the army 47-52
and the RAF 52-56
and the RN 56-61
Defence budget
accuracy of published figures for 169
compared to NATO allies 179
findings of comparative investigation into 183-184
impact of alliance membership on 175-184
impact of Northern Ireland on 174-175
impact of nuclear expenditure on 174
its size 1, 43, 157-159, 167-202
Defence Committee, of House of Commons 40, 126
Nott's evidence to, on procurement policy 94
Defence Export Sales Organisation 84
Defence Export Services Organisation 138
Defence industrial base 1, 4, 88
and redefinition on Europe-wide basis 97
and value for money 162
begging the question? 134
suggested evolution of 94
Defence industrial employment 14, 135, 146
Defence policy
and academic analysts 31, 38
the outsider's perspective 11
Defence research and development (*see also* Research and development)
expenditure, nuclear and conventional 14
Defence Research Establishments 144, 146, 148, 150
Defence review 19, 39-42, 162-163, 166
Defence-contracting industry
aerospace 123-126
competitiveness 16
electronics 123-126
employment in 16
its overseas customers 82, 83
its share of MOD orders 115-116, 151
price arrangements with MOD 103

protection 43, 157
shipbuilding 123-126
size 148
value of R & D performed by 151, 153-155
Destler, I. M. 9
Destroyers 64, 65
Development-industrial circle
and protection 132
and secrecy 117
as an ally of the RN 82
exports 118, 141
friends at court 132-133
gives MOD value for money? 116-128
hesitantly tackled by Nott 86, 92-97
privatisation as protection 136
protected indirectly by Allies' policies 133
protected through global policy 137-144
protection, its cost 152
reaction to declaratory circle attacks 98-102
Development-industrial level 2, 5, 14, 26, 31, 42, 64
and the Nott review 86
inconsistencies within and without 151
Development-industrial/operational alliance 138-141, 144
delivers less than promised 144
Dillon, Mick ix, 32

East of Suez (*see also* Out of area) 35, 65, 78, 80-82, 159
Eberle, Admiral Sir James vii
Edmonds, Martin ix, 33
Ellis, Sir Ronald 84
Equity
and the common defence 175-177
European Fighter Aircraft 112-114

F-111 28, 80
Falklands war 3, 23, 41, 42, 45, 48, 64, 65, 79, 80, 82, 85, 100, 135, 162, 163
and declaratory/operational inconsistencies 84
British materiel shortages in 48
Fearless HMS 64
Five Power Defence Arrangements 67, 77
Flexible response vii, 131
and declaratory strategy 46-47
'Floaters' 84
France
and levels of defence policy 161
and the Independent European Programme Group 98
defence budget 168
its defence budget under volunteer recruitment 171
its navy 43, 65
its official figures for defence spending 169
its procurement policy compared favourably with Britain's 101
out of area policy 182
protectionism towards its defence-contracting industries 161
Free riding 2
and the nature of alliances 196-202
as a determinant of contributions to the common defence 177-181
by NATO as explanation for high British defence budgets 188-196
Frigates 64

GEC 17, 100, 103, 121, 122, 131, 142
Germany 70, 73
and European Fighter Aircraft 113
as arms importer 161
as possible partner in tank production 107
defence budget 168
its defence budget under volunteer recruitment 171

its navy 43, 65
Glamorgan HMS 80
Global policy 183
and defence-contracting industry 123, 153, 161
its cost 159
its success 161
Global role
and the Royal Navy 66
Gooch, John ix
Greece 76
Greenwood, David 36, 37, 53
Gretton, Captain M. P. RN 58
Grove, Eric J. 64, 78
Gulf (Persian/Arabian) 67, 70, 71, 73, 75, 83, 123, 141, 143, 152, 159, 161
as outlet for British arms exports 123, 137-138
British military presence in 75-76
Gulf war 3, 42, 45, 48, 49, 165
British ammunition shortages in 50-52

Hall, Donald 101
HARM air-to-surface missile
preferred by RAF 105, 139
Harrier 81
Hartley, Keith 128, 129, 135, 181
Hayr, General Sir Kenneth 51
Healey, Denis 41, 80, 163, 166
Heath, Edward 78
Hermes HMS 64, 81
Heseltine, Michael 98, 102
Hill Norton, Admiral Lord 28
Hine, Air Chief Marshal Sir Patrick 52
Hunting Engineering 99

IISS 94, 97, 98, 111, 170, 189
Illustrious HMS 64
Imperial role
and British armed forces 35, 67
Independent European Programme Group (IEPG) 97
International Institute for Strategic Studies *see* IISS
International Military Services 138
Inter-service rivalry 28
Intrepid HMS 64
Invincible HMS 58, 63, 80, 81
Iran 67, 68, 74, 143
Iraq 42, 52, 71
Italy 168

Japan 70
Jordan 71, 74, 126
Juste retour 111-113
defined 98

Kennedy, Gavin 180
King, Tom 107, 165, 166
Kuwait 71
balances held in Britain 72
financial assets 73
oil revenue 71

Labour government 157, 168
and out of area 67
Labour Party
and nuclear weapons 20
Lance tactical nuclear missile 79
Lanchester, F. W. 198
Laurent, Paul 33, 34
Lee, John 95, 96
Lehman strategy 57-59
Levels of defence policy 2, 5, 8, 14, 33, 159, 167
inconsistencies between 38, 42, 164
who operates at which level 16
Licence production of defence equipment 153
Lippmann, Walter 33
Little, Richard 13, 31, 32
Lucas Industries 105

Mangold, Peter 72
Marconi 99, 101, 103, 104
Maritime Strategy *see* Lehman strategy
Mason, Roy 122

McGowan, J. 84, 138, 142
McGrew, Tony 34
Military industrial complex 31, 137, 163
Moore, Major General Sir Jeremy 48
Mottram, Richard 40
Mullens, Major General A. R. G. 106
Multinational collaboration (weapons production) 97-98, 110-115, 164
as a spur to efficiency 128-129
less British participation in than meets the eye 103
SP70 48
Multiple Launch Rocket System 48

National Audit Office 154
NATO 76
British membership of expensive 158
defence budget statistics 185-188
genuine cooperation in procurement within 111
NATO NFR90 frigate 114
Naval dockyards 65
Nimrod maritime patrol aircraft 103, 140
Nitze, Paul 12, 38
as precursor of Ball and Snow 9
Northern Ireland 158
Nott defence review 24, 34, 40, 41, 45, 47, 52, 69, 89, 125, 130
its nuclear weakness 79, 85
its probable impact on the RN's budget 85
reviewed 62-65
Nott, John ix, 40, 111, 159, 163, 166
Nuclear strategic force 20, 22
expenditure 14

Oil 70, 87
North Sea 71
participation agreements 72
Oil production, international 87
Oil revenue, international 87
Olson, Mancur 179, 188
Oman 73, 76, 83, 99
and the global role 67-76
People's Liberation Army of 73
Popular front for the liberation of 73
secret war in 3, 42, 45, 73-75
Sultan Qaboos of, installed by Britain 74
Sultan Said bin Taimur of, deposed by Britain 74
Onslow, Cranley 40
Operational level of British defence policy 27, 31, 41, 42, 64, 162
failure at 41
in the Gulf 71-78
Operational level of policy 2, 5, 14
'Options for Change' 39, 165, 166
Oswald, Admiral Sir Julian 56, 60
Out of area (*see also* East of Suez) 64-66, 68, 80, 160, 161
impact on defence budget of 181-183, 189-196

Parkes, Wing Commander Alan 52, 60
Parsons, Sir Anthony 142
Perkins, Major General Kenneth 74, 99, 139
Plessey 17, 84, 121, 122, 138
Polaris 20, 79
Price, Don K. 10
Pringle, Sir Charles 99
Prior, Lord 101
Prisoner's dilemma 196
Procurement Executive 132, 144, 148
and research and development 144-145, 150
Procurement of arms by MOD
broken down by source 108-116
value for money of 89, 151, 160

Qatar
oil income of 72

Recruitment method 158, 171
impact on NATO defence budget

comparisons 172-177
Research and development
and industrial competitiveness 146-148
and multinational projects 112
and the defence sector 144-151
funding of by MOD 120
transferability of from defence sector 149
who pays whom for what 153-156
Rolls Royce 17, 99, 107, 121, 130, 136
Royal College for Defence Studies 3
Royal Marines 64
Royal Navy
and flexible response viii
and global policy 159
and the protection of trade viii
cost of, attributable to global policy 159
its quarrel with the RAF over East of Suez 28
its rationale 1, 157
its strength 1, 43
torpedo requirement 104
Royal Ordnance 51, 105, 136
Royal United Services Institute 3, 4, 56

Sanders, David 34
Sandler, Todd 181
Sandys, Duncan 41, 157, 166
Saudi Arabia 50, 57, 73-75, 83, 140
as arms importer 141
Sen, Guatem 125, 129
Shafer, D. M. 8
Shah of Iran 67, 75, 83, 126
Sheffield HMS 79
Sikorsky 102
Single Integrated Operational Plan (SIOP) 8, 42
SIPRI 123, 170
British arms exports estimated by 118
Skybolt 20
Smart, Ian 67
Smith, Michael 32
Smith, Ron 181
Smiths Industries 99, 132
Snow, Donald 1, 5, 8, 31, 32, 166
Soviet Union 25, 52
SP70 gun 48
Spain
and European Fighter Aircraft 113
Spear, Joanna 140
Spearfish torpedo 104-140
Special Air Service
in Oman 74
Stewart, Moray 116, 117, 119, 121, 125, 130
Stockholm International Peace Research Institute *see* SIPRI
Stoneman, Paul 147
Sweden 178
Switzerland 178

Thatcher, Margaret 102
'The Way Forward' 24, 47
Tigerfish torpedo 104
Tinker, Lt David RN 80
Tooze, Roger 35
Tornado 28, 55, 105, 140
as collaborative project 114
sold to Saudi Arabia 140
Trefgarne, Lord 107
Trident 20, 79, 108, 109, 174
Trident C4 20
Trident D5 20
Turkey 76

United Arab Emirates 72, 73, 75
financial assets 73
oil revenue 71
United Scientific Holdings 117
United States' counterinsurgency policy 8

Vickers 106, 107, 131
Vulcan bombers 67

Walker, Air Vice Marshal 53-55
Warren, Kenneth 95, 99, 132
Watson, Sir Philip 99
Weinstock, Lord 100
Westland affair 102
Wilson, Harold 78
Wilson, I. P. 96, 132, 133, 145
Woods, John E. 36, 65, 81, 82
Wyllie, James 36, 37, 67

Yemen 73